LONDON CALLING NORTH POLE

LONDON CALLING
NORTH POLE

by

H. J. GISKES

Former Chief of German Military Counter-Espionage
in Holland, Belgium and Northern France

E P B M
ECHO POINT BOOKS & MEDIA, LLC

Published by Echo Point Books & Media
Brattleboro, Vermont
www.EchoPointBooks.com

ISBN: 978-1-62654-183-2

Cover design by
Echo Point Books & Media

Editorial and proofreading assistance by Ian Straus,
Echo Point Books & Media

Printed and bound in the United States of America

CONTENTS

PART ONE: SUMMER 1941

THE OFFICER on duty at the Swalmen Frontier Crossing Post, a stoutish major, examined my pass closely and handed it back to me with thanks. He had stepped out of the open door of the Watch Building as my dust-covered sports car drew up sharply in front of the gate which barred its onward passage. He was clearly only giving me his personal attention out of boredom at being stationed in such a dull place, and a single staff officer who drove himself evidently aroused his interest.

"Pleasant journey, Herr Major!" and at his wave the red and white gate rose before me. My parting words were lost beneath the roar of the exhaust and the singing of the tyres on the asphalt road which stretched straight as a ribbon through the frontier woods like a tunnel into the infinite. The good forest air gave my engine a boost on this hot August afternoon, which well suited my urge for speed.

I had reported a week previously at The Hague. The personnel section of the OKW[1] Foreign Intelligence Department was, at least in our rather disrespectful view, much addicted to filling appointments purely on "grounds of unsuitability", and some months previously it had decided to give me a change of air from Paris. Tangier, Athens or Kharkov had been suggested, the latter probably because I didn't know a single word of Russian and had never felt altogether happy anywhere eastward of the Elbe or the Oder. . . .

My director in Paris, the wise Kapitän M——, who was on excellent terms with Admiral Canaris, had in the end telegraphed personally to "High C", as we called the Chief among ourselves, to inform him that I was really a "Westerner", whereupon I was promptly transferred to The Hague to take over command of Section IIIF[1] of Ast[1]-Netherlands.

[1] For this and subsequent abbreviations see Glossary, p. 207.

7

I cannot say that I was very enthusiastic in having to give up my duties and connections in Paris. My section, IIIC2 Ast-Paris, had been operating since the Occupation in May 1940 and had had some success in solving a series of sensational espionage cases in which officials of the American Embassy in Paris and former French General Staff officers had been mixed up. It seemed to me there was more important work still to be done there. What was I going to find to do in Holland, where Reichskommissar Seyss-Inquart was "drawing the blood-related Dutch people back into the bosom of Germania" with a civil government and a tremendous demand for Security Police and SD? In France the situation had at least been perfectly straightforward. The Occupation had developed from the campaign, and the C.-in-C. alone was responsible for the security required by military necessity. . . .

The black banks of cloud with their golden edges had drawn together over the Maas valley, and a downpour of rain accompanied the first flash of lightning as I reached the broad road to Weerd beyond the Maas bridge in Roermond. Knowing that I was overdue at The Hague, I didn't waste time by putting up the hood, but crouched over the wheel and let the stream of water fly past me, like my thoughts of my own homeland by the Rhine, a little way back across the frontier, to which I had just been making a "black" visit.

Eindhoven—Breda—the Moordijk and Maas bridges. The sentries were the single visible reminder of Occupation and war —I was spared the sight of the dreadfully shattered centre of Rotterdam—and then The Hague.

From a short stay there a good two years previously I had formed a picture of The Hague as of an elderly, well-connected maiden lady who likes her creature comforts and still delights in dressing attractively for the benefit of her many friends and admirers. Though spots of field grey now showed on her bright summer costume, she had not lost any of her aristocratic charm.

"A good man of the old school" was my immediate impression as I took a seat opposite the Chief of Ast-Netherlands, to whom I had just reported.

"And what were your duties in Paris?"

"IIIC2, Herr Oberst, with special responsibility for maintaining certain IIIF contacts."

He had bent his small head forward, to hear better, and at this he raised his eyebrows in disapproval.

"Berlin knows quite well that we need a thoroughly experienced IIIF specialist here urgently, and I sincerely hope that you will fill the bill. You will need a day or two to bring yourself up to date with the IIIF situation, both enemy and our own, that is to say in so far as we know it. But make no mistake: I must confess that we are simply feeling our way in the dark as regards any English activity that may be going on here. Abwehr III in Berlin is demanding the closest watch, and Bentivegni and Rohleder were far from flattering after their last visit. That is why your predecessor is leaving. Of course, it may well all be a false alarm—perhaps the Sicherheitspolizei are doing some empire-building, or maybe the Party and the SS are cooking up something for Berlin. Do I need to warn you against these gentlemen? Here in Holland the Sicherheitspolizei and the SD are much more mistrustful and less easily satisfied than elsewhere on account of the local political set-up. General Christiansen is a good man who does his best, but he too is under the Nazi thumb, perhaps without realising it, and should any differences arise you cannot rely on him. As regards the military side, thank the Lord that General Schwabedissen has things well in hand, otherwise I should hate to think what would happen if we had to fight again. A very good fellow into the bargain. We had him and his staff over here in the mess the other day . . ." and Oberst Hofwald forgot his troubles as his mind went back over that last magnificent guest-night.

It was not until much later that I realised what a clever tactician and diplomatic tight-rope walker the Oberst was, the type the Admiral liked so much to use in exposed positions and which helped him to keep the Abwehr ship afloat until 1944. In fact, he gladly gave us a free hand when things went well, and I often had sage advice from him when our "second front" against the Security Police seemed once more to be getting out of hand.

As my car reached the boulevards of Scheveningen daylight had almost gone and I saw before me the unending stretch of the North Sea. I stood and looked at it for a while before entering the headquarters to find my room. Why on earth doesn't the reception clerk ask me whether I am travelling by air or by sea? Oh—of course, I'm still in uniform and there's a war on. As I looked out to sea the war had quite slipped my mind.

The monotonous sound of the surf echoed from the walls of the great white building and emphasised the silence which followed the last march played by the orchestra down in the lounge: ". . . for we are marching, yes we are marching, against England!" The sea front was dark and empty, and nothing was visible except the never-ending ride of the white horses. Clouds drove across the wide, star-pointed sky above the dark no-man's-land of the North Sea, which both divided and connected the fighting fronts, and it came to me, standing there, how water had always been Holland's element and history. For countless ages it had eaten into her flat coasts, but her people had always mastered it and triumphed over the most formidable difficulties. They had made their way by water north, west and south to distant lands and written a famous page in history. Was the next change to come by water too?

I was now facing my own problem, to peer westwards and discover what secret enemy activity was taking place beneath those stars, on those dark waters and in the air above them— activity of an enemy famous for his long experience and unexcelled in his skill at the conduct of underground warfare. We had had a whole series of instructive lessons the previous year in France, Norway and Greece, which had shown me clearly what it might mean to face the experienced toughness of the British Secret Service in combination with an élite of Dutch volunteers willing to risk their lives.

Had it actually come to this? There were stories going around about secret landings from the air and the sea, about radio-links with London and secret journeyings to Stockholm, Berne and Madrid. Were they just rumours—or was there something in them?

The whole of the first day I was shown round the widespread network of Ast-Netherlands in the suburbs of Scheveningen, where we also had our small, comfortable mess. I saw how well the German military detachments had, after a year of occupation, begun to enter into the local pattern of life, to penetrate the heart of the local government and to differ from the national pattern only by the outward and visible signs of military organisation. I realised too how, just as Paris had given its own "Ast" the stamp of its cool intellectual international character, so now, in the summer of 1941, the influence of the broad, placid and unchanging Dutch way of life was reflected in the comfortable, bourgeois atmosphere of Ast-Netherlands.

The "Sealion" roared no longer. The bubble of the projected invasion against England had burst, but was that not proof enough in itself that we had nothing more to fear from the islanders whom we had chased from the Continent?

And were not our panzer armies rolling every day deeper into Russia? Who would think of stopping them when "the greatest general of all time" had ordered them to conquer Moscow, the Urals and the Caucasus . . . ?

The surf muttered on the beach outside—or was it the roll of far-off gunfire? "The German armies will drain away their hearts' blood into the icy plains of Russia, and after two years we shall never see them again." Those words had been spoken by Canaris in the Supreme War Council before the attack on Russia began, and I felt that I was reading them now like a writing on the wall.

The office of IIIF was conveniently situated in a small well-furnished house in the Hoogeweg in Scheveningen. A heavy wrought-iron grille and a wide stretch of front garden gave the necessary seclusion from the little-frequented street bordered by its row of lime trees. The other side of the street was not built up: there meadows stretched away until they merged gradually into Scheveningen wood. The whole could be easily watched from inside the house and every passer-by or vehicle at once seen. Security from unwanted observation was one of the first principles of our Service, and for this reason we had a back entrance, hidden from the street, which led both into the garden round the Ast and into the mess garden. By this arrangement undisturbed entry and egress were ensured for ourselves and for our visitors. The premises next to and behind the building were occupied by the offices of the Naval C.-in-C. Netherlands. There was thus no chance here for the tricks so favoured by espionage and counter-espionage agencies in all countries, by which suspected headquarters are placed under continuous observation and all the occupants and visitors either photographed or filmed. Incidentally, Scheveningen had something of a history in this respect. I had myself seen a film, produced by the German military counter-espionage authorities before the war, in which the entire staff, associates and visitors of the British espionage organisation operating against Germany since 1935 and based on Scheveningen appeared. A couple of cold-blooded sportsmen had taken a film, quite undisturbed, through the port-hole of a large canal barge which from time to time spent days or even weeks secured alongside a jetty not thirty yards from the street in which the British Secret Service headquarters stood. Unfortunately it was only a silent film, but the captions recorded with complete accuracy the names, cover-names, assigned duties, activities and contacts of every single one of the involuntary film-stars. It scarcely needs saying that the British agents in Germany who operated from this headquarters had rather a warm reception, unless, of course, they preferred to get the intelligence required of them direct from the German counter-espionage authorities, against a decent sum of money to be paid by the recipients—naturally! Section IIID of the Abwehr Department in Berlin was employed solely in providing intelligence material which would confuse the enemy, and for supplying it to IIIF agencies when asked for. This "official" trading of false or misleading intelligence was very lucrative, and it helped a lot before the war when the German counter-espionage organisation used to run short of cash!

The British Secret Service probably played just the same sort of game, but with the war things had got a good deal more serious. Now, when a mishap occurred through mistake or negligence it was not paid for only in gold but in blood; and as the struggle grew bitterer and the range of secret operations increased, the responsibility for success or failure often became unbearably heavy.

As I have said, the Hoogeweg seemed to be well guarded against direct observation. But this was merely a beginning: the next consideration was by no means reassuring. Among the personnel running the "Citadel" I could find only one man on whom, in character and professional ability, I could completely rely, particularly when our "second front" against the SIPO and SD was borne in mind. Apart from this one man,

Oberleutnant (later Hauptmann) Wurr, and the NCO interpreter Kup (known as Willy)—an experienced fellow, but a proper rolling-stone—there was scarcely anyone in the "crew" who fitted into my scheme of things. Wurr was some years older than myself, a greying, somewhat rheumaticky man who combined a balanced judgment, great experience and a wide knowledge of men with professional qualities of iron, despite a choleric temperament. Like myself, he had been through the whole beastliness of the First World War as a young infantryman, and if things ever became hot he and his deadly revolver were a sure shield.

"Officers and Abwehr personnel will be in the Chief's office at 10 for a conference!" I heard Wurr announcing outside. I had been able to run over the situation from the reports of individual specialists—as I expected, there was not much to build on—and it was high time to establish the course of action which we would employ from now on.

"Please sit down, gentlemen."

Eight men, some in plain clothes, others in uniform, lowered themselves into the red-leather office upholstery or basket-chairs brought in from the sun-room close by.

"Herr Major," reported Wurr, "I have just arranged for both the heads of the Radio D/F group of Intelligence Section IX in Holland to be here at 11 a.m. Oberleutnant Heinrich from the Radio Observation Office of the ORPO has also asked for an urgent appointment."

"Thanks, Wurr. Fix Heinrich for 11.30. Have you any idea what the listeners are on to?"

"No, Herr Major; they just said it was urgent."

"Good. And now, gentlemen, to go on to what I have to say to you. In the past few days you have individually given me a fairly clear picture of affairs. I will sum up my impressions as follows.

"Enemy. No active individuals or organisations are at present known with certainty to be engaged in secret service, espionage or sabotage in Holland. There are, however, a few indications of such activity. The proof that forces exist in this country which are in contact with London lies in the recent unfortunate case of the Security Police on Sneeker Meer, where the party lying in wait for an enemy seaplane engaged in running agents was shot up by the aircraft itself. London must already be aware of SIPO's activities, and the enemy now holds the initiative.

"In Spain we have identified and uncovered, partly by photography, a clearing-house for Dutch espionage reports to London, and I am at this moment awaiting the full report. From it we are hoping to discover whether these Dutch groups are working independently or on directions from London. Please keep clear in your minds the fundamental difference between these two possibilities. The countering of independent activity in Holland is first and foremost the duty of the Security Police. They are in contact for this purpose with the Field Police, the Military Patrol Service and Ast Section IIIC2 according to circumstances. If on the other hand the Allied Secret Service in London is directing sabotage or espionage activities, then IIIF, the Military Counter-espionage Service, comes into operation. The same obviously applies to all cases of radio communication with the enemy.

"As regards clandestine movements by sea, there have been numerous unconfirmed reports which it is impossible to check. The BBC material transmitted by Radio Orange also gives an indication of the movement of secret agents. We shall take special steps to pick up these threads in conjunction with the Naval Abwehr and the coast-watching service. I hope quite soon to get some detailed information from the Radio D/F authorities.

"Our own position. This is that we have no leads into enemy secret organisations through our IIIF 'V' personnel either in Holland or abroad. There are in the interior a few loose threads which may possibly lead to such organisations. All these must be closely investigated, and if necessary money can be spent freely.

"As regards the suitability of our agents the 'V' organisation does not appear to me to be altogether successful. For instance, it is quite wrong for us to employ known NSB members. If the slightest suspicion were aroused, these men would be unmasked at once. And such men, when in paid Abwehr employment, are prone to misuse their position to deal with individuals whom they dislike on political grounds. I intend to remove 'V' men of this type as soon as opportunity offers. And, *meine Herren*, let me make it absolutely clear that we shall in no circumstances whatsoever have any dealings with party or political matters.

"My principle will be to use a few first-class agents who may be highly paid and must be really suited to the work. Too large a 'V'-man organisation brings danger from leaks and chatter. I shall give my directions later on as regards your personal relationships with 'V' personnel, particularly women, and my standing orders concerning contacts by 'V' personnel with this office. Please study them carefully.

"Finally, the 'provocation' method as advocated by Hauptmann Kleebacher is not to be employed. I am well aware that any decent Dutchman can be led astray by a skilled and conscienceless provocateur, to get him to do something for the Allied Secret Service and so render him liable for punishment by the Occupation authorities. I forbid such methods absolutely, and this is in accordance with the directive of Abwehr Section IIIF in Berlin. Please consider yourselves warned in this respect.

"Now another question of principle. Our activities in pursuit of our object of uncovering enemy secret activity may lead to the discovery of punishable criminal offences. This, however, in contradistinction to SIPO, is not our business. Our one aim and purpose is to discover the secret plans and communications of the London Intelligence Service in such a way that we can mislead and thwart its sinister projects. By this means we can acquire important information for our High Command, and in order to accomplish it we may have to allow enemy agents and organisations to operate for a while undisturbed, provided that we have identified them properly and can adequately control them. The processes of arrest, house-search, requisition and every other kind of police activity are the job of SIPO. We are not criminal experts or detectives, neither are we magistrates or executioners. Our duty as Abwehr officers is to prevent criminal acts, not to punish them. Remember, too, that while a man may be sentenced lawfully, morally he may still stand head and shoulders above his judges.

"In cases of armed resistance, make good and timely use of your weapons—but remember again that a defenceless opponent is no longer an enemy but a man whose only crime is to love his country the same as you or I do.

"That will be all for today, except to announce that the Chief, Admiral Canaris, is inspecting this Ast towards the end of August. I hope to have an opportunity of discussing with him the IIIF situation in Holland. Herr Wurr, please remain behind. Thank you, gentlemen."

Chairs scraped, and after an exchange of salutes the heavy door swung to behind the last man.

"Cigarette?"

"No, thanks, Herr Major, I'm a cigar-smoker. But please try one of mine," and Wurr pulled out a vast cigar-case.

"Thanks very much."

We smoked for a little in silence. The evening before, we had been interrupted in the middle of an interesting discussion in the mess by the entry of Oberst Hofwald. Wurr had made good use of his few weeks at The Hague to penetrate quite astonishingly behind the scenes. The devil only knew where he had got all the "Court gossip" about the Reichskommissar, the C.-in-C., and SS Führer Rauter. He was really very well informed.

We got now to discussing two local society ladies, by the name of P——, who played a not inconsiderable part at the "Court" of the Reichskommissar.

"Is it confirmed," I asked, "that P—— himself is in London working for the Dutch Government in exile?"

"The Foreign Censorship Department in Paris has it from letters sent to Frau P—— by friends in Madrid. What's more, I think that Frau P—— is probably working for SIPO or SD in connection with the Reichskommissar and his circle. As both these ladies are as catholic in their methods as they are in their choice of friends, they are no doubt excellently informed. The short-sighted and careless use of their contacts is not much of an advertisement for SD. But in any case they are not Allied secret agents."

"I should rather say 'Not yet'. Both by their contacts and their temperament both ladies would seem to be thoroughly well suited. . . ."

Two field-grey open army cars drew up sharply outside, and just as if we were looking through a bioscope we saw two identical young officers jump simultaneously out of two identical doors. The effect was so remarkable that both Wurr and I looked at one another and laughed.

"If they fix their positions as well as they drive . . ." said Wurr.

"Ask them in," I told the girl as she opened the door to announce their arrival.

Salutes—introductions—handshakes followed. These two D/F unit leaders had the air of well-trained pedigree gun-dogs, for they had ten years of hard soldiering and specialist activity behind them. What a contrast to the tired lot of sceptical or cynical reservists who had just left! Still, there was no time for such reflections—the maps were already laid out before us.

Oberleutnant O——, whom I had known in Paris, spoke for the two of them, briefly and clearly. He and his interception, D/F and decyphering teams had made themselves talked about "upstairs". He had now been working for six weeks on a close D/F search for agents' radio transmitters in Holland. Some weeks previously the German interception stations in Norway, Poland and southern France had reported short-wave links with England just established in the Netherlands area. The position had not yet been "fixed" accurately, but the type of traffic betokened an agent network. The irregular times of communication, short traffic-periods, prompt answers to calls and other indications all told the same story.

"Our latest D/F bearings indicate the presence of two agent-transmitters to the northward of the big rivers. On this map I have put down the close-range D/F results. According to these, a station with call-sign UBX is operating within the triangle Utrecht–Zeist–Amersfoort. The set transmits irregularly on 6677 and 7787 k/cs. Its working routine is not fully established, but we have confirmed about five periods a week. The period of transmission is six to nine minutes. Up to the present we have intercepted fourteen signals sufficiently accurately to make it possible to decypher them, provided we can find the key. The cryptographic department of the Army Intelligence Radio Section in Berlin has broken its teeth on it so far. We suspect a Dutch method of encyphering. If only we can make a successful capture we ought to be able to lay our hands on sufficient material to solve the problem.

"A second transmitter with call-sign TBO has just recently started operating in the Delft–Gouda–Nordwijk area, probably quite close to The Hague. We know no more than that at the moment. Leutnant R——will be working from today on observation and close-range D/F of this station. I am putting my own men on to close-range D/F of UBX in the Utrecht area.

"As the ORPO listening post in Scheveningen comes under you, Herr Major, in so far as it deals with agent traffic, I ask that its observations should be merged with my own both retrospectively and from now on. If it is intended to use IIIF personnel, once we get close-range bearings, I should be grateful for the closest co-operation in all details."

Wurr and I had listened closely, and Wurr had noted down the most important points.

"When do you think you will get close enough to UBX to make a seizure possible?" I asked. "This could mean a great deal more than the elimination of one agent, since a successful capture would make available material which could clear up what has been going on round here for so long. As you know, the Abwehr headquarters in Berlin believes that Holland is of much more interest to the secret agencies in London than the Reichs-Kommissar appears to think. Why should the Allied Secret Services have a less favourable field of operations here, for example, than they have in France? Because the Dutch have meanwhile rediscovered their German hearts? They didn't exactly greet us with open arms in 1940. These Goebbels fairy-tales aren't even fit for schoolboys! I believe the English know perfectly well there are no longer so much as six German divisions along the whole Atlantic Front. Our job, therefore, is to find out his methods and intentions so as to meet them with our own. That's why I am asking about a possible date of seizure."

"If the operator carries on working as he has done and without disturbance from us we could manage it in a fortnight from now. If, however, he grows suspicious or changes his position for some other reason no date can be guaranteed," replied Oberleutnant O——, drawing a map from his brief-case. "Here is a comprehensive report from Military Radio Intelligence Berlin giving the latest counter-agent results in all the war areas. It contains the latest information about the enemy's methods of secret radio operations obtained from the entire German interception and D/F organisation. It comes out monthly and I will see that a copy is passed to you direct.

The French section of this number has some interesting observations about rapid position and frequency changes, as well
as giving the call-sign changes of stations operating in France
since March 1941. This enemy habit complicates our job
immensely, but it is clearly necessitated by the heavy losses
he has sustained up to the present. Nevertheless, if UBX stays
where he is for another fortnight I think I can promise you
we shall get him."

"Can you leave this report with me?"

"Of course, Herr Major, if you will give me a receipt and
send it back personally."

Wurr brought me a red Top Secret receipt form, which I signed.

"I hope you two gentlemen will lunch with us?"

"Very sorry, Herr Major, but we have a twenty-four-hour
schedule, which depends entirely on UBX and TBO. When
they are off the map we will gladly visit your mess."

Men after my own heart! By the time I had got back to
my room after seeing them out the two cars had already
disappeared.

Shortly after noon there called to see me a wiry man of
middle height in the uniform of the Ordnungspolizei, whose
manner and speech betokened the old type of highly trained
Security policeman. He gave me an account of his duties in
charge of the Radio Interception Service of the Ordnungspolizei, known for short as the FuB station of the ORPO, and
did not fail to make it clear that he considered himself as
coming directly under me for all agent interception work.
He added that he had nothing to do with the Security Police,
and that he reported only to me and to his superiors in Berlin.
He was very well up in the Abwehr side of agent radio work
and had already been in communication with Berlin about
transmitters UBX and TBO.

Although, quite clearly, all assistance was welcome, I was
concerned at the number of departments which were involved
in this work. Here once again we could see the cursed duplication without which nothing could be done under Nazi rule!
From top to bottom and round every corner one ran up
against this so-called "competition", which made continual
friction unavoidable, yet which was proclaimed as the essential
condition for State service within the "thousand-year Reich".

I decided that I should have to play politics. Duplication must on this occasion be made to operate to our advantage, and rivalry might help towards a quick solution of the problem. I therefore decided to aim at his vanity. "I am very pleased with your report, Leutnant Heinrichs. It is quite remarkable how quick you have been in gathering information about the stations which have just opened up. Let's have plenty more of it. You and your men can soon do the job, if necessary without the help of the military D/F detachment. How strong is your section?"

"All told, about twenty-five men," he beamed. "Five or six of these are on continuous twenty-four-hour listening watch. Although our actual area of responsibility is limited to Holland, we take on D/F interception work for other countries, particularly England. We are also responsible for jamming enemy radio-propaganda transmissions."

He spoke for a little while longer, and I had a clear impression that he was "military-minded" and wasted no love on the SIPO. This, and the fact that he and his men were superlatively good at their job, would be very useful to us later on, although I had no idea of this as yet. At the moment I clearly had him on my side, however, and Wurr grunted with approval when Heinrichs had made his departure, with a bow which would not have disgraced a puppet-theatre.

Since shortly after my arrival my uniform had hung in a cupboard awaiting some official or warlike occasion before again seeing the light of day. Abwehr officers were permitted to wear plain clothes when necessary for the proper performance of their duties, but the military conscience of our superiors came into operation when too liberal use was made of this privilege. We had to watch matters in this respect, particularly in the Occupied countries. The few active-service officers who had remained over from peacetime and had a background training were already rather like needles in the haystack of new entries. Actually a few reserve officers had had short periods of Abwehr training before the war, but there was as yet no question of a properly trained and completely efficient organisation. Wherever opportunity had offered practical experience in addition to the normal training excellent results had been obtained. In any event, and apart from service considerations, plain clothes were much more appropriate to the inviting beach of Scheveningen, to which in this summer of 1941 thousands of civilians flocked daily and which was hardly improved by its flavour of military cloth—quite the contrary!

At the Grand Hotel, however, service dress attracted no attention. It was fuller than in any peacetime season, and the pre-war international kaleidoscope had given place to a brilliant patchwork of national uniforms, from navy-blue, through bluish-grey and field-grey to the well-represented and beloved "sovereign brown" of the Party, with lots of gold. The "mixed" quality of both sexes of visitors to the hotel, moreover, left little to the imagination. . . .

For four weeks I passed as a holiday-maker, until I was able to move into a room in the office-building from which I could control all parts of the IIIF organisation. In these first weeks the crying need for good new men never gave me a moment's peace. Since 1940 the Abwehr had extended from the North Cape to Bordeaux, from Finland to Athens, and more recently to Russia and North Africa, and by now the situation as regards suitable trained personnel had become critical. We searched everywhere for people with the right character qualifications, a knowledge of foreign languages and with sufficient common sense. They must not in any case be "party-boys", and those with any leanings towards the SD were also out of the running. During 1940 I had been able to collect a "team" in my Paris office which was remarkable for its mixture of military and civil elements. Businessmen and musicians, aristocrats and sailors, scientists, foreign legionaries, adventurers and "bourgeois" were all represented—all of German nationality—the rest being soldiers of various ranks. But this last differentiation meant little, since from my point of view the personnel in my office did not just consist of officers, NCOs and men. Much more important was the distinction between "Abwehr specialists" and "others". The other ranks were partly officer-trainees, and it was nothing out of the ordinary to dress up a leading-seaman as a lieutenant-commander, complete with papers and decorations, and send him to play this role in different offices and among various authorities, as he was fully capable of doing. Although very independent in thought and behaviour, men trained and led in this way soon discovered the need for *esprit de corps* and willing obedience, which was not just confined to the giving and taking of orders. The ambition to make a success of a mission counted with them much more than obedience to strict instructions, and recognition by the Chief meant more than tangible rewards.

I had had to leave them all in Paris in the care of my successor, but after a fortnight there came a cry for help. The worthy Major Feder, who had taken over from me, was not going down too well with a set of men who respected a chief for his capacity rather than for his rank. Things were going badly, and the "team" were inclined to pay small attention to his instructions, so would I like to arrange for the transfer of a few men whom I could use in Holland?

I went to see Oberst Hofwald. "Please send me to Paris for three days, Herr Oberst!"

"You've just come from there. What makes you want to go back?"

"I haven't a single man here, Herr Oberst, whom I can put on to the penetration of enemy courier or agent networks with any hope of success, and there isn't much doubt that these networks stretch from Paris and Brussels to Berne and Madrid. But there are good men in Paris who are sitting twiddling their thumbs because they can't get on with Major Feder. As I shall have some vacancies here shortly, I should like to have them transferred."

"Vacancies? How do you manage that?"

"Because, Herr Oberst, in the next day or two I shall be asking for the transfer of a number of redundant personnel. In my opinion, any side of Abwehr work other than IIIF is more suitable as a home for perfumed officers whose only wish is to draw their pay, with or without aiguillettes."

Hofwald looked at me sharply. Was that aimed at him, the Commanding Officer of the Ast? Or was it that the speed of my actions didn't suit the careful diplomat? Had he been taken by surprise, hurt in his pride? He loathed anything which did not accord with his cultivated and well-arranged conception of things. Any decision taken without consideration of all relevant factors or reached by any kind of loose thinking aroused his mistrust.

He remained silent for an uncomfortably long time. "Put off your trip for the time being, Giskes," he said at last, "until the Chief has been. We don't know what will come of his visit, and I should in any case like to have you handy for the next few days. Then you can go off to Paris. Let us go a little more deeply into this question of transfers now. I don't want the newcomers simply to repeat the old failings."

"I ask for your confidence in this, Herr Oberst. I need a small staff here, but an expert one, with no 'passengers'. If we are really going to build a wall against the unknown threat which hangs over us our material cannot be too reliable nor carefully chosen. Personally, I don't believe that they are asleep in London, just waiting, in fact, for Hitler to eat them up. They've paid too dearly already for that. All the trump cards are in their hands for fighting a secret war behind our backs, and they will soon start to play them. They have had plenty of time to make the necessary preparations, and I don't want to find myself with empty hands when things start moving."

Hofwald, friendly as always, definitely thought that I had painted too dark a picture, but I could see that I had made an impression. Ten minutes later I was through on the direct line to Major Feder at the Abwehr headquarters, Paris, and had fixed my visit for the first week in September.

In the keen cold of an early September morning the dew ran in broad streams down my windscreen and dripped on me from the old trees which overhang the "Napoleon" road between Breda and Antwerp. The frontier-post at Wüstwezel lay behind me, and now the trees were flying past to left and right like palings. At this rate I should be in Paris in time for lunch at the Abwehr headquarters, at the Hotel Lutétia. I had no urgent problems this time, and driving myself like this was a welcome relaxation after the demands of the previous few days, which had left me with a set of impressions still needing to be worked out. The two-day visit of the Chief, Admiral Canaris, had ended the previous evening with an official dinner at the Oude Doelen, which had gone off very well. The party of about a dozen had included the Chief of the Security Police and representatives of the Wehrmacht staffs. When the guests had broken up into groups after dinner,

Canaris had seized the opportunity to have a long heart-to-heart with Hofwald and Harster, then, true to his usual form, had left at ten o'clock. This talk gave us all food for discussion until midnight, in which each of the four men in our small circle participated sarcastically or destructively according to his individual style.

Canaris had emphasised the necessity for us to work closely with the SIPO and had warned us: "Mind you see I'm not bothered in Berlin with your jealousies." It was clear to the initiated that this warning meant that he was anxious about the future of the Abwehr, which he looked on as a counter-balance to the totalitarian ambitions of the Reich Security Headquarters. He could do his work only if the relations between his Abwehr stations and the corresponding SIPO and SD authorities remained peaceful and without fundamental differences of opinion. One conflict of views about the field of operations of the Abwehr and Reich Intelligence Services had already resulted in acceptance of the RSHQ viewpoint, an outcome made inevitable by the strong position occupied by the RSHQ and the party-leanings of Keitel, head of the Wehrmacht Oberkommando.

We often discussed the personality of Canaris, with his frequently contradictory instructions and sometimes incomprehensible intentions. This shortish man with his mass of grey hair must have cut a fine figure on the bridge of a ship, and his large, blue eyes, wide awake and intelligent, betokened an officer of the old Imperial Navy. He seized every opportunity he could to get out of his admiral's uniform into plain clothes. On the morning of his arrival he had inspected the line of officers waiting to meet him and had name after name given to him as he shook each by the hand, adding a nod of the head and a screwing-up of the wrinkles round his eyes for the old hands. When he spoke he avoided set speeches and contented himself with a few quick and often humorous remarks which made each of his points crystal-clear in a couple of sentences. In conversation with two or three people his mask of reserve soon dropped, and then he revealed himself as the possessor of great knowledge of men's minds. His pleasant, soft voice rang with the force of an inner conviction which ten successful years of Secret Service had given him. Before the war an important English weekly newspaper, referring to Canaris's Greek name and origin, said: "Hitler had made the most oriental of his officers head of his military Secret Service." It was a shrewd thrust at the Nazi race-complex, but the author had hit the nail on the head about Canaris, probably without realising it. The combination of impenetrability, intelligence and cunning must have seemed to his enemies the more subtle by reason of its rarity among the senior officers of the Wehrmacht. But no one knew him completely, not even those closest to him. The ends for which he strove have been discussed over and over again since the end of the war, but all we knew at the time was what he did *not* want, namely, any infringement whatsoever of the unwritten laws of humanity. In this respect he was never untrue to his decisive rejection of the false values which had arisen in the Germany of the Third Reich, even though, as an experienced soldier, he knew well that war makes its own laws.

He had listened approvingly to my rather sweeping views about the IIIF position in the Netherlands, and asked me for my impressions during the few weeks I had spent there. "Make it as short as you can," he said; "we all know that one's predecessors never do things quite right." Everybody grinned with appreciation and we were all glad to be rid of the necessity of having to discuss unpleasant matters.

This time Canaris had stayed once again with his old friend, Kapitän Richard Patzig of Wassenaar. Up till 1937 this man of nearly seventy, known as "Uncle Richard", had been one of the leading figures of the OKW Abwehr headquarters in Berlin, where he was at that time head of the Counter-espionage Section. Patzig, who was astonishingly well preserved, had lived since 1938 in Wassenaar, looked after by his factotum, known as "Auntie Lena", his housekeeper, secretary and the sharer of all his service secrets. Before the war, in the guise of a representative of the German railways, he had spun out the threads of his web from this Eldorado of holidaymakers and idlers against the English Secret Service operating in Holland. Station "P" was noted on the maps in Berlin against many Top Secret matters taken from his reports. Since May 1940 and the coming into being of Ast-Netherlands, Station "P" had swung into the orbit of the official organisation. But it still remained directly responsible to Berlin, and none of us really knew all that old Uncle Richard was up to.

His friendship with Canaris went back over nearly forty years. At that time Patzig had been Canaris's officer-instructor when the latter was a cadet in the Imperial Navy. After the First World War they had gone together into Secret Service work, which Canaris had taken up when he ran the Naval Espionage Service in Madrid in 1916 against the Allies of those days in the Mediterranean. There were countless legends about his exploits in the First World War. When his cruiser, the *Dresden*, was sunk in 1915 he escaped from internment in Chile and made his way back to Germany in 1916 right through the enemy controls, disguised as a Chilean merchant. In 1917 he was taken prisoner by the Italians, sentenced to death, but escaped back to Spain in his usual adventurous fashion.

My car was running well. Brussels, Maubeuge, Le Cateau, Laon, Soissons—so many memories of the First World War. Was it perhaps because this rich rolling hill-country with its ancient farmsteads and half-empty villages was in such contrast to the carefully tended and over-cultivated Dutch countryside that these memories were so sharp? Clear sunlight lay over the Ile de France, and the delicate pale-blue tints of the southern horizon led my thoughts on towards the object of my journey—Paris.

The check by the German military patrol at the Port St. Denis was perfunctory. German civilians in cars with French markings were a very common occurrence. . . . I wanted to spend this visit in Paris as a civilian and to stay in my old quarters, where the landlady knew me as Herr Gerhardts, *monsieur le docteur*. For this reason I had put French number-plates on my car before leaving The Hague. I had special written authority to make such changes as convenient, a pass issued by the Chief of Staff of the Army in France certifying that I was entitled to travel at will with German or French number-plates, in or out of uniform. Furthermore I was authorised to cross the frontiers of Germany and the Occupied countries at any time or place, and to enter all restricted areas and military installations. Persons accompanying me required no passes, provided that I vouched for them. This comprehensive document was stamped and signed by the Chiefs of the Patrol Services in France, Belgium and Holland. Normally I used the ordinary route-orders and travel documents in order not to make unnecessary use of these privileges, but when, as often happened in later years, I had to act quickly and take individuals of various nationalities across frontiers and through restricted areas without having the time for the customary formalities, my pass came into its own. It was not, of course, made out by name, but by description, beginning "The individual answering to the above description. . . ." For this reason I always had the choice of the various cover-names at my disposal for use with other types of pass. The unavoidable and multifarious contacts with the local population resulting from the long period of Occupation could not be allowed to provide any loopholes through which the enemy Secret Service might penetrate the German counter-espionage organisation.

It was like a home-coming to be in Paris again, the city, where, unlike anywhere else in the world, the stranger is offered a second homeland. What had it been like in May 1940? The last sounds of fighting had scarcely died away before Paris had herself conquered the breathless victors with astonishing ease and speed. We had come unasked and forced ourselves on her, so no one could say we were welcome, but nothing could stop us from falling victims to her irresistible charm.

I passed through the crowded dining-room of the Hotel Lutétia to my old place, exchanging greetings as I went. Each one lunching at these small tables was part of "the works"— the Abwehr. Officers, secretaries, assistants. The men were nearly all in their forties or fifties, the women scarcely yet in their thirties. A colourful mixture of bright summer frocks, formal or sporting plain clothes and well-fitting uniforms.

"Papa" Feder's large bald head was full of the troubles he had had with my bunch of rascals after my departure, and he could hardly wait for coffee to be served in his office before starting in. His armchair creaked ominously beneath the weight of this heavy man, sweating with excitement as he poured out his complaints against my friends and former assistants. It was clear I should have to pour some oil on this raging sea of outraged officialdom and offended dignity, and the tension was eased slightly when I asked him a question or two about the progress of some important contacts which I had handed over to him. Nothing of much importance had occurred meanwhile, but he seized delightedly on this opportunity of riding his particular hobby-horse before a professional colleague.

After a while it proved no longer difficult to persuade him of the desirability of breaking up the "crew". "Arno" and "Oswald" were to come to me in Holland, and the remainder, finally absorbed in Section IIIF, Ast-Paris. The agreement was sealed by an evening's celebration with my men.

Rrrrr—Rrrr—Rrrr—— The shrill bell of the direct telephone from the Hotel Lutétia to my quarters in the Avenue Gabriel broke into my Sunday afternoon nap. I replied with the cover-name for the flat.

"Duty Officer Lutétia speaking. Is that Dr. Gerhardts?"

"Speaking."

"There is an urgent Top Secret signal for you from The Hague. Will you please collect it personally?"

"Thank you. I'll come at once."

On the way I racked my brains as to what its contents might be. It must be something special and urgent, otherwise Wurr would simply have asked me to ring him in Scheveningen. Had the D/F people found their man? The fortnight's grace which Oberleutnant O—— had given the operator of UBX had by now elapsed.

I had guessed correctly. The message ran: *UBX seized at 0800 today. Operator and assistant captured. Codes and extensive espionage material secured intact. . . . When will you be back? Wurr. Oberleutnant.*

I thought quickly. This was our first important stroke in Holland, and if properly exploited it could have far-reaching consequences. What a fool I was to be enjoying myself here in Paris when I ought to be in Scheveningen! An hour later I was roaring northwards again through Le Bourget, the Lutétia duty officer having told Scheveningen to expect me at 2200.

The seizure had been a complete success, thanks to accurate short-range bearings obtained by Oberleutnant O——. It was a remarkable fact that the D/F listening-gear was still recording call-signs from the transmitter at the moment when the Security Police had already forced their way into the house under suspicion, and this fact had caused Wurr, who was close by, to enter and search a summer-house which stood on the same piece of ground.

Downstairs, a large room—empty. Nothing suspicious here. Quick, upstairs! In a half-darkened room Wurr sees before him two men, who spring to their feet in surprise. He quickly grasps the situation, and his "hands up" holds the men rigid beside their set, enabling the D/F troops who are following to secure the set with all written material and much gear besides. I had the full report from him the same evening. The captured espionage material, a pile of typewritten sheets, consisted of about forty serially numbered reports prefixed AC, and hundreds of individual items.

"Good. We'll examine this a bit more closely tomorrow. Are the names of the arrested men known?"

"No report yet from the Security Police," said Wurr. "Leutnant Heinrichs has possession of the transmitter and all the gear. His men are now reconstructing the code. Heinrichs will be reporting here at 0900 tomorrow."

"Has the Abwehr headquarters Berlin been informed?"

"I have already informed Berlin in your name."

We separated with a warm handshake. The new day just dawning would tell us more.

It had to be confessed that the prospects of "playing back" this radio set looked poor. The operation known as "playing back" consisted of continuing to work a captured set in the service of our own counter-espionage, in order to conceal the fact of capture from the enemy Secret Service and to leave it under the impression that its agent was still operating safely and undisturbed. This playing-back was a fundamental principle of German military counter-espionage, and every other consideration and motive had to give way when it became possible for a radio-link with the enemy to be established which the latter might regard as absolutely genuine. Messages emanating from an enemy Secret Service and received by a radio set under our control could give us valuable information about enemy intentions. Every task given to agents, every question passed to them and every exchange of messages was a milestone along the road towards the objective of counter-espionage—to penetrate the heart of the enemy

Secret Service. In our present position this was, generally speaking, the only chance of obtaining quick, reliable and complete intelligence. Yet while the heart and brain of the Western Allied Powers were centred in London, physical penetration of the island stronghold by our agents was attended with such risks and such loss of valuable time as to be virtually impracticable.

In this respect the enemy was distinctly better placed. It was much more feasible for him to penetrate Germany and the occupied countries through Switzerland, Sweden and Spain, not to speak of the support which enemy agents received from the inhabitants of occupied countries and from the mass of foreign workers in Germany. On the other hand, having to accept the fact that no such opportunities existed for us in England, we applied ourselves all the more thoroughly to the device of playing-back enemy radio sets—a game we had already played in various ways in 1941 in France, at first with good results; and although we had not discovered how it was that the Secret Service in London had seen through our game so soon this did not stop us from continuing to develop our tactical methods.

From the report of Leutnant Heinrichs it appeared that the code used by the UBX operator contained certain special features whose meaning and object could only be guessed. We had succeeded in deciphering many of the messages from UBX which had been intercepted previously, but we had as yet insufficient knowledge to enable us to play the set back. The code had been identified as a Dutch naval type.

I had not yet received any report from the SIPO giving the results of the interrogation of the operator, which might have added to our information, and I waited for this impatiently. Long delay had already made a play-back difficult. An operator could always miss one or two routine transmitting periods, but any longer interval would arouse suspicion "over there", and I accordingly sent Wurr over to the SIPO for information.

At midday he returned to say that the SIPO had nothing useful as yet to give us—they had embarked on the normal police questionnaire, beginning with "Parents" and "Place of birth". Wurr had run into a complete lack of understanding of our requirements, and it had been conveyed to him that it would be preferable if the Abwehr could keep its nose out of police affairs. Previous experience had not encouraged me to expect anything different. At that time the SIPO, generally speaking, would not take the military aspects of a case into consideration so long as they could make a major criminal affair out of the capture of an agent which might have wider ramifications and lead to further arrests. It was much more in their line to work up a case like this into a gigantic show trial. How satisfactory to have such an opportunity of "liquidating internal resistance" and, above all, of checking up whether this or that name taken from the list of their political opponents could be connected with the case. In France conditions were more favourable for Abwehr interests, since in that country only the Secret Field Police were entrusted with powers of arrest and interrogation—a force under the orders of the Military Commander-in-Chief which co-operated excellently in all matters touching the interests of the Wehrmacht.

I outlined the situation shortly to Hofwald, who looked doubtful.

"You must become better acquainted with the local situation, my friend," he told me. "I have it on good authority that in the SIPO we are dealing with a man whose chief concern is unsavoury political affairs and police activities. He is called Schreieder or some such name. See if you can win him over. Perhaps he might actually co-operate with you, provided the idea is not frowned on from up above."

Cheap advice—but none the less sound.

Wurr went round in the afternoon to the SIPO, with the result that the Herr Kriminalrat Schreieder expressed his willingness to pay me a visit the following morning and to acquaint me with the results of his investigations. Meanwhile to the Abwehr headquarters Berlin's enquiry whether the transmitter could be played-back I replied non-committally.

A small, almost bald man with a heavy round head, in the uniform of an SS Sturmbannführer, entered my office next morning and extended a flabby, well-manicured little hand. With a series of small bows he preceded me into the sun-room, where he placed himself in a chair and crossed his short legs. During the exchange of conventional platitudes which followed I had leisure to observe him more closely. His age was difficult to judge—perhaps forty. Slightly protruding, rat-like eyes gave lift to the pasty face, and the nose betrayed the delights of the bottle. The whole well-fed man exuded joviality, his slightly provincial accents emphasising the note of southern warmth, as though he was immensely pleased to have found in me an entirely unexpected and beloved old friend. He radiated the well-known benevolence of certain criminal investigators, before which Edgar Wallace's murderers are supposed to dissolve into tears.

So this was the man whom, in the opinion of Oberst Hofwald, I should have to interest in military affairs. To judge from his intimate manner and the friendly *"Lieber Kamerad Giskes"* with which he continually addressed me, this should not prove too difficult. One thing—I could be quite sure that he had already had my background investigated and that he was fully conversant with both my private and political pedigree. I went straight to the point and put it to him that it was unfortunate that forty-eight hours should have elapsed since the arrest of this agent without any investigation being made into the possibility of playing-back—this being the only side of the affair in which I was interested.

"Lieber Kamerad Giskes" he began, "please rest assured that I shall do everything for you which can be reconciled with police requirements. The operator is a tough fellow in spite of his youth and is not to be broken down at the first onslaught. All that we have got out of him so far is that he is a Dutch naval cadet working for an espionage network operated by the Dutch Admiral Fürstner in London. He alleges that he was put ashore from an MTB two months ago on the Dutch coast, and that he has not since moved his radio-transmitter from Bilthoven. So far he remains silent about the organisation which has been collecting and decyphering the captured messages, and the identity of his assistant has also not yet been revealed. He has not come from England, and seems to be a courier for the espionage organisation. He has spun us one or two yarns about how he allegedly got hold of his material, on which I am checking up. Meanwhile I have brought you an extract of the interrogation report so far, which may be useful to you, and will keep you informed of whatever else emerges."

There was nothing for it but to assure him of my conviction as to the correctness of his approach and my appreciation of his frankness towards me.

I received subsequently a series of excerpts from the interrogation of this agent which told me virtually nothing except that his real name was, apparently, Zomer. The final result of this case only became known to me nine months later, when Zomer and a large number of Dutch who were accused of espionage appeared before a German military court and were condemned to death. From my experience in this case I learnt the lesson that, cost what it may, I must try, given another opportunity, to get the radio-operator into my own hands.

The start of my work in Holland did not seem to be favoured by fortune, and for the next few months one reverse followed another. A seizure was planned of agent-radio set TBO, operating in The Hague, but this was a complete failure. Probably warned of the capture of UBX in Bilthoven, he had concealed himself much more efficiently. Close-range D/F bearings indicated that the transmitter must be in a block near the Staats-Spoor station at The Hague, all the flats in which had separate entrances. A Dutch "V"-man from IIIF was given the duty of reading the electric meters in the block during the transmission period, disguised as an electricity official. Under this pretext, he was to take out the fuses of each flat in turn for a short interval, by which means it was hoped to establish in which part of the block the transmitter lay, through the sudden interruption of the signals.

The officer-in-charge, who was watching the doors from a house opposite, knew exactly where the "meter-reader" was at any given moment, and at first all went perfectly. The agent was transmitting, the D/F group listening, and the "meter-reader" had entered the first of the suspected blocks. A minute afterwards the transmitter was suddenly interrupted, and when the "meter-reader" emerged two minutes later he was followed closely by a young man with a box under his arm, who mounted a bicycle and rode away. When the D/F men burst into the house and found the nest empty they could only assume that this man had, in fact, been the operator. It was discovered later that the moment the "meter-reader" had entered the house the landlord's daughter had warned the agent, and it was a two-minute job to pull out the plugs and decamp. The small number of men watching the house had aided his escape, and by now he would be miles away.

This incident produced an immediate storm from Berlin, with many subsequent rude remarks. Since we had not planned an actual capture for this unfortunate day, the SIPO had naturally not been called in, and complaints about this reached me from all sides. But it could have gone wrong in any case. For if we had told the SIPO what we intended to do with our "meter-reader" they would have turned up with the usual mass of cars and men, and the operator, having been warned, would doubtless not have touched his key at all, which would have led to a similar failure of the operation. I consoled myself with the thought that station TBO was no longer working and that the ether above the Netherlands was "clean" once more.

Under the code name "Hauskapelle" every IIIF section had attached to it a party of paid agents who were experts at their job, and who were always available for special duty. There were, when I arrived at Scheveningen, four young NSB men belonging to the "Hauskapelle" under the leadership of a shifty, fat, bald-headed, sixty-year-old "Boss", who offered their services during the first week of my arrival. The "Boss" and his gang roused my heartiest suspicions. All that I had heard about their previous activity had been pretty sinister, and I planned to get rid of the lot as soon as I could. Among other instructions for the guidance of the "Boss" I laid it down that he was to be the only one with authority to come to the Hoogeweg. Hitherto the whole party had gone in and out as they pleased. A few days after the issue of these strict instructions I overheard Dutch being spoken outside in the hall—it was one of the myrmidons—and that fairly settled it. My next interview with the "Boss" could have been easily audible through the padded double-doors of my office. It had at least the benefit of making my orders clearly understood, until there took place shortly afterwards a bloody incident which brought the existence of the "Hauskapelle" to a sudden and dramatic end.

Sunny October weather, calm and warm, lay over Scheveningen and gave no hint of the unpleasantness to come. The "Boss" had been to me with a report that his men were in contact with an enemy espionage group in Brabant, probably in the town of Breda. His men had offered to take espionage-material for this group along the courier-route via Brussels and Paris into Spain. His answers to my detailed questions were unsatisfactory, particularly as to whether the contact had been established from our side or theirs, but he promised to give me fuller information the next day. I advised him not to rush his further investigations, since haste could only result in suspicion being aroused, and told him to keep me informed of developments. Shortly after noon on the following day he reappeared with the news that his men would be meeting representatives of the espionage group the same afternoon. Three of them, who had evidently gained the confidence of their opposite numbers, would be meeting them in a flat rented by the latter in Haarlem, while the fourth man would remain on guard outside the house during the interview.

I did not attribute much importance to these events. Not one of the numerous investigations of alleged spies carried out hitherto had led any further, and as I was only too glad to see the "Boss" in my office as little as possible it did not worry me when he failed to appear for two days after this visit. A surprise was, however, in store! On the morning of the third day, a tattered, unshaven, hollow-cheeked individual staggered into the office, who showed hardly a trace of the self-confident phlegmatic "Boss" whom I knew. He collapsed into a chair in the sun-room, where, half-whimpering, he began to stammer out disconnected sentences, from which I made out the words "Murder—prison—all dead—SIPO" but little else.

It took some while to sort out approximately what had happened, and this was serious enough. The three men of the "Hauskapelle" had been shot to bits by their visitors. They had scarcely sat down at a table when the leader had drawn his automatic and opened fire. His companions had also fired till their magazines were empty, while only one of our men had managed to shoot, and his shots went wide. Result—on our side one dead and two severely wounded. The criminals had escaped, although the man on guard outside had given the alarm at once. When they arrived on the scene the SIPO could only confirm this unfortunate state of affairs, but they seized the "Boss" when he turned up at the flat an hour after the shooting, so as to have at least one arrested man to show for it. They had locked him up and interrogated him until the morning of the third day, contemptuously ignoring his protestations that he worked for the German Abwehr and his requests for them to refer to his Chief at the Hoogeweg.

And now the wretched creature sat here in a state of complete collapse. Had he just realised that war means shooting and death, and that no mercy is shown in the shooting affrays of the underground struggle? Had he only just found out that, when it comes to the point, the secret struggle between espionage and counter-espionage does not shrink from murder and assassination? Or was it simply the anxiety for his comfortable "soft job" which had finished him so utterly? And what had led him to take sides against his own people but his own urge to earn dirty money through informing and betrayal? As I looked at him I could only feel anger and disgust at the dreary figure before me—at the war, at the Service in which I worked, and at myself.

The "Hauskapelle" was broken up, and all that was left of it dismissed. The bloody affair had revealed that there existed in Holland organised lawlessness which would defend itself by any means available. The police investigations pointed towards Breda and on the track of a Dutchman who had been an instructor at the military school there. He had got away in time, and we learnt later in the war that he had made his way to London and taken over a leading position in the Dutch Secret Service or "BBO" (Bureau Bijzondere Opdrachten).

The mess-up over TBO and the bloody failure at Haarlem fully justified an immediate reorganisation of our "V"-man system. It had become clear that comprehensive security precautions would have to be taken before we got close enough to an agent's transmitter for him to suspect that something unusual was going on. It would therefore be much better to put our people into civilian clothes and engage them in harmless occupations rather than use them as a mass of armed and uniformed men. From now on we could no longer count on achieving the kind of surprise which we had managed in the case of UBX in Bilthoven. They had had their warning, and now knew that by means of our D/F bearings we could ferret them out of the darkest of corners. They would probably now use in Holland the same devilish means of defending themselves which operators in France had already employed to cover their retreat—explosive charges in doors and in the radio sets, half-emptied brandy bottles with poisoned contents and other nastinesses. There seemed to be no limits to ingenious assault.

Two rounds of the duel in Holland had now been fought and a two-to-nil result recorded against us. It was true that no sets were now in operation, but I was quite confident of their early reappearance and took the necessary precautions accordingly both at headquarters and in the "V"-man organisation. It was impossible to keep a gang like the "Hauskapelle" on a sufficiently taut rein. Each individual "V"-man needed his own Abwehr officer to "run" him, to give him his orders, supervise the results and pass them on to headquarters. From now on the office must be guarded like a bank and only specially authorised individuals allowed to enter it. The concentric outer rings of Abwehr officers and "V"-men, with their informers and unwitting sources, represented not only the instruments of our probing into the enemy's front but a shield and filter against undesirable penetration of our own organisation from without. Only a small number of "V"-men could be allowed access into the heart of our organisation beyond the inner ring of Abwehr officers, and this only in exceptional circumstances, while contact between individual "V"-men must be prevented as far as possible and each must only know his own particular Abwehr controller. The whole organisation must be reduced to a small nucleus of first-class collaborators with Abwehr headquarters.

I impressed this more and more on my staff during the next few weeks.

". . . I must rely on your ingenuity to find the right answer when you have understood the general sense of my ideas. I cannot obviously lay down standing orders and you know as well as I do that there is no training manual for service in IIIF. You will take the proper precautions only if you remember that any contact with a 'V'-man may be under observation by the enemy, and that the 'V'-man himself may suddenly for some reason start working for the other side. The preliminary contacts with a 'V'-man are like an advance into no-man's-land, where surprise by the enemy is possible at any moment. Use initiative in making the small deceptions which are necessary in the long run for success and for your protection. Change your name, your appearance, your cars and their numbers, and above all change your meeting-place as frequently as possible. Keep your meetings as short as you can. I will not permit regular contacts to be arranged for reasons of convenience or thoughtlessness. I require original ideas which can help to solve our problems. I do not require a large organisation, and the number of our collaborators must stand in inverse ratio to the effectiveness of our work. We can only achieve success if we confine ourselves to our primary object, the penetration of the enemy's secret organisation. Should we discover any gap or joint in the enemy's armour I will concentrate all our resources to ensure penetration at that spot. . . ."

The office had had a welcome addition since September. "Arno" and "Oswald" had arrived from Paris. I detached Arno to Amsterdam at once, with the object of picking up the threads of the courier lines which ran from Holland via Brussels and Paris to Switzerland and Spain. As he spoke French perfectly, he worked from the first days of his arrival in Holland in the guise of a French or Belgian businessman who travelled to and from the south. Oswald remained at headquarters. Hauptmann Kleebacher, a type who was not well suited to our organisation, was recalled to Germany at my request, together with two others. We hoped to get replacements, but it was not until the spring of 1942 that four more officers arrived, and later these justified themselves nobly. The Wehrmacht D/F detachment was withdrawn from Holland at the end of September, as no more suspicious traffic had been heard and they were urgently required elsewhere. Interception watch was left in the hands of Leutnant Heinrichs and his men from the FuB office of the ORPO in Scheveningen. Relations with the Security Police remained cool but correct. I had returned the visit of *lieber kamerad* Schreieder, which produced no better professional result than had our first discussion after the UBX affair. It seemed as if we had been granted an unexpected breathing space, a last pause in which to marshal our strength for the coming struggle.

PART TWO: OPERATION *NORDPOL*

CHAPTER I: PRELUDE

AS ALWAYS in the late afternoon, the vestibule of the Hotel Excelsior, opposite the Potsdamer Station in Berlin, was teeming with humanity. On the way to the lift a good sniff of coffee on a passing tray enticed me towards an unoccupied table in a corner, where I soon forgot my weariness and my previous intention to rest for an hour in my room.

Two days in Berlin, crowded with visits to senior officers in OKW on the Tirpitzufer and meetings with Radio-Abwehr colleagues in the Mathaïkirchplatz had extended me considerably, not to speak of the previous evening's celebration with Rockoll and Bosenberg at the Femina. This evening would set things right, as the train would not be at The Hague before tomorrow morning. "Our armies are driving forward with heavy engagements" the headlines of the evening papers shouted at me. I skimmed through the Wehrmacht communiqué. It required a form of second sight to reconcile its terse phrases with the true facts of the position on the Eastern Front which I had just had at first hand in the OKW. The Russian winter had set in early, and with sudden intensity, which had brought the decisive phase of the advance on Moscow to a halt. During the past week the Russians had made significant territorial gains, and the front had begun to consolidate. Our losses were not yet ascertainable, but they must be enormous. . . .

The lively bustle in the vestibule had increased. What a lot of oriental and Balkan types one saw here in Berlin! Was it they who had brought this curiously oppressive, strained, hectic way of living to the city, which could be felt almost physically after the quiet formalism of The Hague?

"Hullo, Doctor; what brings you here?" said a low, pleasant masculine voice with a slight Viennese accent close beside me. Without turning round, I knew it to be "Freddy", the "Ace" of my Paris contacts, charming and elegant as ever, who in his search for a seat had arrived by chance in my corner. "Well, one certainly has to come to Berlin nowadays, if one wants to see the high-ups," he went on, without explaining whether he meant me or himself. "They relieved me in Paris in September. I've heard all about you, urgent recall and so forth. What was it all about? Oh, of course—Top Secret as usual. But you look fine . . . hardly recognised you. Just look what sea air and plenty of sleep can do."

His coffee had arrived by the time I could get a word past his chatter to ask him a question. In Paris the previous winter I had got to know well this intelligent, splendid-looking scion of an old Austrian family. He was running an industrial concern integrated into the German armament programme, and shortly after I had got to know him he came to see me over a rather delicate matter. He was seeking my protection for a beautiful young Frenchwoman, the bearer of one of the best-known ducal names in France. He had discovered by accident that she had helped a friend of her family, a French General Staff captain serving in the English Secret Service, in his dangerous work, and he knew that the captain's espionage group had already come to the attention of the Abwehr.

What should one do in such circumstances? I made up my mind that a twenty-year-old girl who had become the accidental accomplice of a skilful secret agent would be warned off once and for all by the arrest of her friend and his group. I thought I could probably "square" my immediate superiors if I let her off after taking this action, but it could prove fatal if my helpful attitude became known in unfriendly quarters. Fortunately the incident had passed off without attracting undue attention, and a few months previously my now sworn friend Freddy had married his blonde, blue-eyed "Daisy", having already himself become a German national through the annexation of Austria. They were the happiest, handsomest and most elegant couple in all Paris.

Freddy, as usual, talked vivaciously and continuously. He knew hundreds of people in all the Axis and Allied capitals, travelled frequently, saw a lot and heard everything! During dinner and up to the time of my departure in a sleeper to The Hague his stories threw a clear light upon the dark background of the war-panorama. He had a good knowledge, from long residence there, of the Anglo-Saxon countries and their industry.

"Take care to keep your stomach in good order, Doctor," he said as we parted. "When Hitler starts his war against America we shall soon be getting some pretty indigestible victuals, and unshakeable faith is a poor substitute for bicarbonate of soda."

He was standing outside on the platform, and the light from the carriage was reflected in his dark, lively eyes and on the smooth, dark hair of his small head.

"So long, Freddy. See you in Paris; and my regards to *madame la princesse*."

And so the train moved off.

In the Schauburg the last strains of "Butterfly" had died away, and enthusiastic applause had brought the really superb artists of the Berlin Players several times before the curtain.

I drove back slowly through the cold, starry night to Scheveningen. Nervy and disturbed as I was, I couldn't get the tune of the "longing" motif out of my head. ··· — ··· — ··· — it continued, ··· — V, ··· — *Vertrauen, Verrat, Verbindung*? The leads (*Verbindungen*) across the North Sea which it was my business to create. I had been given this on the Tirpitzufer in Berlin as my most urgent task. A means of lightning communication, of short, rapid, feverishly transmitted radio messages which would spurt out like jets of water, like a shuttle thrown into a secret, invisible loom which would draw tight the net in which must be caught the sinister plans which were being prepared over there on the island against the Wehrmacht.

During the afternoon I had given Oberst Hofwald a detailed report on my visit to Berlin. We had discussed, without illusions or excuses, the Abwehr situation in the West, which had become still more threatening since June 1941. Since the attack on Russia, Soviet sympathisers in Western Europe had become active, whereas they had previously "stood at ease". A network of Russian radio stations had been heard operating since August by the Radio-Abwehr Interception and D/F Departments, and new ones were now appearing in France,

Belgium and Holland each month, like mushrooms from the earth. Two shortwave agent links between Amsterdam and Moscow had been clearly identified, and bearings taken, since the beginning of November. The Communists in Paris, Brussels and Amsterdam had clearly not been wasting their time since the "Treaty of Friendship" of 1939. This widespread activity, which had started up during the summer, plainly had behind it a well-supplied, well-prepared and highly skilled organisation.

This, however, was only one side of a very acute danger. A much more serious probability was that groups of fanatical idealists who had hitherto held themselves back under orders had now found their way to the Allied Secret Service headquarters in London and were thus in a position to join up freely with Resistance groups in the Occupied areas which were ideologically inclined towards Great Britain. These dynamic, revolutionary forces, whose strength was growing "over there" beside the enemy, by whom they were carefully trained and led, would have no compunction in carrying the war behind our lines in all conceivable forms. The great increase of secret radio traffic between London and Occupied and Unoccupied France was a sign which could not be overlooked! When would Holland come into the scheme of things?

I explained to Hofwald what Berlin had in mind. The necessity for the Wehrmacht to put a stop to this invisible and dangerous enemy activity had apparently penetrated to responsible authorities which were not connected with Abwehr or Radio-Abwehr headquarters. The seriousness of the position had, it seemed, caused Himmler's men to instruct the police chiefs in Occupied countries that they should, for the time being, sink their differences and the rivalries between themselves and the Abwehr of the Wehrmacht. If this were to be carried out at lower levels we should have much better prospects than hitherto of "playing back" captured radio operators.

The lights were still burning in the Abwehr mess, with Hofwald and Wurr occupying armchairs in front of the fire. Their heads, sharply outlined in the light of the standard lamp, turned towards the door as I entered the half-lit room. We shook hands and Hofwald poured out a third glass of "cordon rouge" from a bottle in the cooler beside him as I drew a chair up to the pleasant warmth. The mess was Hofwald's pride and joy, and when he felt like it he could be a magnificent host.

"General Schwabedissen wished to be remembered to you, Herr Oberst," I said. "He was sitting with Von Müller and Janssen in the box close beside me. I thanked him in your name for the tickets which were sent us from the C.-in-C.'s office. The show was really first-class, and Mierenholz made a superb 'Butterfly', with a ravishing voice. Seyss, being a Viennese, has a weakness for music and the theatre—how nice it would be if he could be put in charge of that side of things alone! There were two men in the Reichskommissar's box, neither of whom I knew, and, by the way, Madame and Mademoiselle P—— were all by themselves in the box next door. It was rather noticeable in the interval that there was not the usual 'convoy escort' round those two pretty women. Can it be they have fallen out of favour at Court?"

Hofwald lifted his glass towards me. "In that case you could make it a Service affair and take them over," he said, smiling. "A IIIF expert ought to be able to make first-class use of such material."

"Quite so, Herr Oberst. And when one's actual plans have been made this type of 'material' can come in very useful in certain circumstances for a special purpose of short duration. But only according to the rules observed by the Secret Service when making use of clever and pretty women, the rules which suit these women best; that is to say, lay siege, succeed, put on ice, and don't employ a second time if you can help it! The problem of introducing such ladies into longer-term Secret Service work is only for those who have plenty of time on their hands."

Hofwald looked at me thoughtfully, as he always did when he wished to hear more.

"I would naturally make an exception," I continued, "for those cases where a passionate woman, following a personal impulse of great love or deadly hatred, can do astonishing things for a secret mission. If one comes across such a rare bird one is only doing right if, out of human respect, one gives her the chance of proving the strength of her inner driving force. Ordinary adventuresses such as the Mesdames P——, however, are only after the superficial kind of success. They do homage to money alone, and this god provides them with the wherewithal for satisfying their other impulses."

"A good thing they can't hear you," said Hofwald, motioning the waiter to refill our glasses. "Let's rather stick to our iced champagne, which is more satisfying than your cold counsels. Your health, *meine Herren*. Let us console ourselves with the thought that there are other women than the harpies whom our hard-working Giskes has conjured up."

"Herr Oberst, I drink with you to that thought. I should be the last person in the world to withhold my admiration from an attractive woman—provided, of course, I haven't got work on my mind."

Wurr accompanied me part of the way down the Hoogeweg. There had, it seemed, been a report from "Willy" about a meeting with an individual who had offered himself as a "V"-man using the name of Ridderhoff, and who claimed to have got wind of a secret affair which concerned English agents at The Hague. The man was after money, and his story sounded a bit nebulous. "If money is all he wants, then we can do something for him," I said. "However, good night, and thanks for the report."

· · · — · · · — · · · — V, *Verrat! Verbindung?* It worried me still.

The following morning the mail was particularly voluminous. It was put on my desk every morning at nine, together with the map, and guarded by Wurr, with whom I used to discuss the in-letters. Unfortunately it usually consisted of useless paper. Innumerable offices, bureaux, special detachments and liaison staffs had sprung up like dragon's teeth in this first year of the Occupation, and they were still on the increase. Civil Government and Wehrmacht, Party and Police, the armament industry, the war-economy people and the Todt organisation, all with large staffs, had acquired offices, and now made every effort to prove their indispensability by means of vast piles of paper. Duplication, so typical of governmental practice in the Third Reich, had reached undreamt-of dimensions since the beginning of the war, in spite of brave talk of reduction and economy. It was clear to anyone inside it that this must end simply in organised chaos.

Thoughts like this were with me this morning, and I cursed

the growing heap of paper before looking at it more closely. Wurr was sorting the incoming telegrams, which usually held most interest.

"Another addition to the family, Herr Major. Our request for Huntemann for IIIF is approved and his marching orders are on the way to Ast-Copenhagen. Then there is a transfer order for a Major Wiesekötter, now in Ast-Athens, to come here at once. Athens has signalled that he is flying via Rome and Munich."

"Just a moment, Wurr. I know Herr Wiesekötter from Paris. I had him for training last winter. He is a reserve officer, slightly addicted to alcohol. I wonder how he got on at Athens. If he has asked for a transfer it simply means that there's nothing decent left to drink there. Apart from that, he's just our man."

"Then Ast-Wilhelmshaven is asking whether we have any objection if our old 'Hauskapelle' team, which has now recovered, can be used there as 'V'-men. They will be employed in the branch office at Groningen."

"My dear Wurr, you know how much I disliked Berlin's orders that Ast-Wilhelmshaven IIIF should work part-time for Naval Abwehr in Groningen and Friesland, although responsible also to me. I can't alter that. However, I wish them joy of our former cavaliers. But what else is there? Have you Willy's report about meeting the new 'V'-man?"

He searched for a moment, and then drew out a half-sheet of paper, evidently torn from a large notebook, with a few lines of Willy's regular handwriting on it. They didn't convey much . . .

"The Hague. 27th November 1941. Personal for Head of IIIF. Today at 1300 hours second meeting with Ridderhoff, American Hotel, Amsterdam. R. states he is in contact with Dutch reserve officer who works for two English agents operating probably from The Hague. R. needs money. Also asks for protection against the German currency authorities, who have had him in custody charged with diamond-smuggling. Request interview with Chief 1700 hours 28th November at headquarters. Willy."

I read the note through carefully and handed it back to Wurr.

"Make sure Willy fills in the appropriate personality forms and other particulars about the new man directly after our meeting this evening. And give him an agent-number and cover-name at once. The name Ridderhoff must in no circumstances appear in any further reports from Willy. That obviously applies to all our discussions as well. Has anyone other than you and I seen this report since last night?"

"No, Herr Major; it was in the Top Secret safe all night. I placed it there personally myself. The key was put away as usual in a sealed envelope and left with the duty-officer."

"Good, Wurr; I know I can depend on you for such details, but the chance of a success must not be thrown away through neglect of such trifles. The good fortune which has brought us this new man could turn against us tomorrow if we don't preserve absolute secrecy. Whether Willy's report gets us anywhere we can't yet say. But if it proves to be genuine, then I'll ride this horse till its last breath. Arrange matters so that you can attend the meeting this evening, and see that we are on no account disturbed. Good—and now let's put the rest of this paper through the mill."

At my call "Bring him in!" a large, powerfully built man of about forty entered my office behind the girl-clerk, whom she announced as "Unteroffizier Kup". His dress and careful appearance were those of a well-placed official or businessman, and his healthy, open face gave an impression of satisfaction with himself and the world. Not even the most suspicious-minded could mistrust his approachable and winning personality. Kup was a born "contact-man", with an infinite variability of approach to his fellow-men. Nevertheless I still had some reservations about his suitability for our service, although it was true that he had not for some time had an opportunity of demonstrating his native shrewdness and his trained intelligence, which caused him to do the right thing in every set of circumstances. In my opinion he would handle important and delicate matters with circumspection tempered by a healthy self-confidence, and for this reason I now intended to let him run this new contact himself, without putting in an Abwehr officer as intermediary.

His salute and form of greeting illustrated the slight offhandedness which I encouraged in him, as in other of our outside collaborators. This relationship, and the mutual forms of address of "Willy" and "Chief" underlined our special feeling of confidence in one another, besides affording the disguise which was sometimes necessary. We sat down at the table, while Wurr placed Willy's report about Ridderhoff before us. It was already stamped "Top Secret".

"A week ago," Willy began, "I met by chance in Amsterdam an acquaintance whom I had not seen for a long time. His name is Pieters, and all he knows of me is that I visit Amsterdam from time to time on business. He has, of course, no idea that I am a soldier and that I belong to IIIF. Pieters told me that he had been in custody and under investigation for four weeks on suspicion of black-market activities. They could prove nothing and eventually released him. During his period under arrest he had made a number of useful contacts. One man, for example, had told him in confidence that he had reliable and secret means of communication with Paris, which he used for smuggling diamonds. This person's contacts were particularly circumspect and well-disguised by reason of other illegal activities in which they were engaged. I at once pricked up my ears, but let Pieters continue without interruption, merely saying at the end that I should like to get to know this man. I had been on the lookout for such a line, and if I could make use of it all three of us would be the gainers. The day before yesterday Pieters introduced me to this dealer in diamonds. His name is Ridderhoff. We discussed smuggling only, and Ridderhoff gave me a telephone number in Baarn which would find him if I needed him.

"Well, I telephoned yesterday morning and fixed up to meet him at the Carlton Hotel in Amsterdam at 1300. He was there punctually, and soon brought the conversation round to his lines to Paris. He mentioned several times that he was in difficulties over money, and that he would soon have to 'bring off something big'. When I intimated that there were certain German authorities who paid well for information about secret lines of communication he became very interested. This seemed to be particularly important to him as a line of defence against further investigation of his case by the Currency Control. Ridderhoff had emptied a bottle of Bordeaux in the course of a light lunch, and he now began to relate adventurous stories from his life. Opium-smuggling and prevention in the Far East were much to the fore, and he implied that a well-paid job in Secret Service was just his line. If it were to be against England, so much the better.

"I gave him to understand that I might perhaps be able to put him in touch with some such work, but he would first have to convince us that he had the necessary contacts. He then brought out his point about a captain at The Hague and the two English agents. I didn't want to squeeze him dry on the first occasion and felt that a few banknotes would stimulate both his memory and his greed for more."

Willy hesitated and looked at me questioningly.

"When are you seeing this man again?"

"Tomorrow at 1200 in Utrecht, Herr Major."

"Herr Wurr, please make a note of this discussion. Number will be F2087—cover-name 'George'—pay 500 florins. Further enquiries about him to be instituted at once. You, Willy, make him the offer tomorrow, and tell him that there will be a large bonus for quick results. This creature is not concerned with moral issues. But make it abundantly clear that any fairy-tales or the slightest concealment from us will be followed by his immediate arrest. Get close to him and find out all his particulars and habits as accurately as you can. From today you will be relieved of all other duties. When necessary, apply to Oberleutnant Wurr for assistance. And make quite certain of the kind of man George is. What have you discovered so far?"

"He lives in Baarn, Chief, and is allegedly in business. When in drink he speaks a mixture of Spanish, English and Dutch. He is a large, fat, bloated sort of fellow—lame in the left leg—a type you can pick out from a hundred."

"Good. Get a photo—under the pretext of getting him a false pass. You are quite clear, I hope, that we are only concerned in uncovering the agents and their collaborators at The Hague? Keep George strictly on the rails. No new ideas and no side-tracking until that is cleared up. Got it?"

"*Jawohl*, Chief!"

"Report back here as often as you can, but only to Oberleutnant Wurr or myself. Good luck!"

He shook hands and left the room.

In Willy's next report, F2087 stated that the agents at The Hague possessed a radio transmitter. I had already warned Leutnant Heinrichs and the FuB station of this possibility. Telegrams were immediately despatched to Berlin and to Radio-Abwehr with an urgent request for a careful watch by all stations on suspicious traffic between Holland and England. Leutnant Heinrichs put four men on special watch, but the days went by without anything suspicious being reported.

Five or six reports from Willy registered no progress. F2087 was evidently engaged in gaining the closer confidence of the captain, which involved expenditure by him—and by us. After ten days we had not advanced a step. Heinrichs swore that nothing had been transmitted from The Hague, and Radio-Abwehr, Berlin, had asked me some supplementary questions which I could not answer, which did not improve my temper.

On 10th December a report from Willy ran: " . . . source F2087. Agent Two at The Hague is looking out for suitable sites for dropping weapons and sabotage material by parachute. The timing will be arranged with London and a reception committee detailed by him. . . . A widespread organisation is being planned, which will be systematically armed and trained. . . ." I grabbed my red pencil and wrote in the margin: "Go to the North Pole with your stories. There is no radio communication between Holland and England. F2087 has three days in which to clear up this contradiction in terms!"

Although I was fairly certain that F2087's reports about communication were nonsense, I rang up Heinrichs and was rather short with him. "Why isn't your interception watch doing something? I have a report here that a radio-link exists. Dropping-points are being investigated for future parachute operations. If you can't hear them, I shall conclude that your men are hibernating!"

"We haven't heard the slightest thing, Herr Major. We keep continuous watch, and all German interception stations have instructions to watch for this particular traffic. We can only carry on, and I don't personally believe that this set is in fact operating."

"Thanks, Heinrichs. It's possible, of course, that the rascal who gives us these reports is double-crossing us. I will let you know at once if anything fresh turns up."

Furious with it all, I picked up Willy's report. "What do you think, Wurr?"

He shrugged his shoulders. "There seem to be two possibilities. Either F2087 is leading us up the garden path despite his five hundred guilders, which must surely show him we are serious. If that is so I can promise him a warm time. I'm sure, all the same, that Heinrichs and his men are working well and doing all they can to trace this link. On the other hand, it may, of course, be that the transmitter is of a new type. I heard in Berlin that great progress had been made in the design of VHF[1] sets for agents' use. It is quite possible the enemy is using these and our normal interception service can't pick them up. Radio telephony has also to be considered. Our own portable R/T sets can only work over a few kilometres, but how can we tell how far the enemy has improved his?"

"If Willy comes this afternoon, make it quite clear that I shall not in any case wait longer than three days before arranging to have F2087 put under close observation. That would have great disadvantages, as the man will soon realise what is going on—but this question *must* be cleared up."

In the evening Wurr told me that Willy had listened without comment to my remarks on his last report.

"Did you ask him if he wanted to see me again?"

"It didn't get as far as that, Herr Major. We had only been sitting a few minutes before he got up and left. He even left untasted his strong cup of tea, which he can't normally get on without. Your remarks certainly stirred him up."

"Excellent! Then he is in just the right frame of mind to talk to friend F2087. Otherwise, the man may go on thinking he can 'milk' us and retail his knowledge in driblets!"

Two days went by without a sign from Willy. Leutnant Heinrichs came once or twice to our building, but did not ask to see me. He was evidently testing out the atmosphere. At 9 a.m. on the third morning there appeared in my office two smiling men who scarcely bothered to put on their normal official manner—Wurr and Kup.

I took up Willy's report and read: ". . . concerns *Nordpol* [North Pole]. . . . Source F2087." I looked up. Both men wore

[1] Very high frequency.

wide grins. I had no option but to laugh with them. Willy had given the Operation a code name which had a particular point to it, and he had every right to do this, as his present report had removed the last of our doubts. ". . . F2087 had been in close contact with Reserve Captain v.d. Berg at The Hague since 12th December. Berg had accepted F2087's offer to take care of the transfer of certain material which will be delivered from England. Drops are not expected for the time being, as the radio link with England is not in operation. The set appears to be defective, and efforts are being made to repair it. . . ."

At last we had a harmless explanation of the misunderstanding which had been giving us sleepless nights. But that wasn't everything. More important still was the delay caused to the work of the enemy group through the missing radio-link. This delay should give us an opportunity of bringing F2087 so closely into contact with the leadership of the enemy group that we should be kept informed in good time of all their moves. The leading individuals could now be investigated at greater length, for F2087 would be closely concerned in the internal build-up of the organisation. By this means the enemy could carry on developing his group undisturbed though under continuous observation by us, without the necessity for an early seizure. I worked out this idea in detail with Wurr and Willy, but I could not guess what results would come later from the use of IIIF principles in Operation *Nordpol*.

Events in the following weeks took their expected course, and by the turn of the year F2087's penetration had succeeded so well that reports about Captain v.d. Berg brought in valuable clues about illegal activities quite apart from those touching the two agents who had been dropped by parachute.

Early in January 1942 things started moving. From v.d. Berg F2087 learnt that a plan to take three men by MTB from the coast at Scheveningen to England was in course of preparation. The MTB would approach as close as possible to the shore and pick up the three men at the second dyke to the northward of Scheveningen pier. The day of the operation would be communicated to the group at The Hague at 2100 hours over the BBC transmitter Radio Orange. If one evening the "Wilhelmus-Lied" were played over Radio Orange that would indicate that the MTB had sailed to pick up the men between 2100 and 0100 hours the same night. . . .

This report raised a number of problems. First, how had the arrangements for this operation been made? Second, how were the men or their material or knowledge so important that it was worth while to risk an MTB to fetch them? Or was it just a trial to see if a future link could be set up using this method? For us it was all important that now at last F2087's reports could be put to the test. If they stood this test they would be given correspondingly more weight in the future. We had thus both to watch the enemy's attempt closely, but without arousing suspicion, and at the same time we had to prevent it if that were possible. A close watch on the approaches to the beach could only be maintained by the regular patrols of the VGAK, a Customs Preventive Service. For this reason I informed the local inspector as far as was necessary, but I naturally didn't tell him about the rendezvous signal, only that I would inform him of the time at which his men should keep a particularly sharp look-out.

The subsidiary problem—to prevent the operation from taking place—was more complicated. Weapons might have to be employed, as it could only be a British or Dutch warship with which we were dealing. I therefore made preparations to mount two heavy machine-guns in the middle of the pier, which would be manned by an experienced Feldwebel from the Abwehr Guard detachment. Leutnant Heinrichs was given instructions to report the playing of the "Wilhelmus-Lied" over Radio Orange at once to me or to the duty officer at headquarters, using the code word "National Anthem".

A week passed before anything happened. I was sitting one Sunday evening in my room reading, and thinking of anything else but that our first night operation in Holland was to take place this very night. At 2115 the duty officer rang up and gave me the code words "National Anthem". This signal set all our preparations in motion.

It was a cold, stormy night, and heavy clouds driving from the west continually covered the quarter moon, which had just risen, when I drove to the pier at 2300 to see if all was in readiness. In order to avoid the slightest slip-up, the heavy gate which shut the pier at night had been closed as usual.

So I had to clamber over the gate, like the machine-gun's crew before me, shortly after passing a patrol of the VGAK. Visibility was good, and the intermittent moonlight was sufficient for us to watch the sea to the northward as far as the third dyke. For security reasons the machine-gun crew had been provided with Very lights and had orders to open fire on any boat which approached the shore. The time went slowly by, and we passed night-glasses round so as always to have a fresh pair of eyes keeping a look-out. Since we had intercepted the correct signal at 2100, in accordance with F2087's information, it was reasonable to suppose that the second half of his report, which concerned the arrival of the MTB, would follow. But nothing happened.

I left the pier at 1 a.m., but the machine-gun's crew remained behind with orders to stay on watch until dawn. About 2 a.m., however, I arrived at the headquarters of the Coast Patrol, to be met with the surprising news that at about midnight three men had been arrested by one of their patrols. They had been seen climbing into an old bunker at the north end of the promenade. There was little doubt in my mind that these were the three men referred to in F2087's report, although the arrested men explained their breaking of the curfew by saying that they had been to a party and needed "a breath of fresh air". It was hardly possible to prove that they had intended to go in an MTB to England. The slightest sign that we knew more might easily endanger the confidence which they placed in F2087. For this reason I simply asked the Security Police, through the Commandant of the Patrol Service, to take over "three men who had been arrested on the shore in suspicious circumstances".

In a brief report to the Security Police I sketched out the preliminary moves which had led to the arrest of the three men and asked for an enquiry to be made into their background. This aroused lively indignation. The fact that the whole affair had passed off successfully and in accordance with my intentions didn't prevent the Security Police from protesting sharply to Oberst Hofwald. Reasons—insufficient preliminary information to the SIPO about my plans. This new tension between the SIPO and my office had the unfortunate consequence that I was not given the results of the police inquiry into the "beach affair", so that we were left still uncertain how arrangements had been made with London for the signal to be passed by means of the "Wilhelmus-Lied".

Shortly afterwards we heard from Leutnant Heinrichs that an English radio-operator had been arrested by the SIPO. He had been working only a short time, and his name was v.d. Reyden. This man should have been "reversed" and his set played-back by the SIPO. But for reasons which were carefully kept from us nothing came of this great project, and we found out subsequently that a whole number of collaborators of the "beach walkers" had been arrested.

Although we were ourselves in the dark concerning the details of the v.d. Reyden case, what was quite certain was that among the staff of the IVE section of the SIPO at The Hague there was not a single man who was competent to play-back a radio transmitter. As I heard soon afterwards from Heinrichs, this truth had penetrated meanwhile to Kriminalrat Schreieder, and successful efforts at mediation by Heinrichs did eventually result in the police tacitly deciding to leave the next case of an arrested radio-operator in the hands of IIIF, should any possibility of playing the set back arise.

We now concentrated on a closer investigation of the group led by Captain v.d. Berg which was running the radio operators, in the hope that a slice of luck combined with skilful preparation for subsequent action might provide us with sufficient material for a "play back" to be initiated. F2087 was very busy in these January weeks. I now took his reports much more seriously, and two events which occurred at this time again confirmed the reliability of his information. These events had nothing directly to do with the plans of our enemy in London, but so far as I remember they provided valuable confirmation of the genuineness of F2087's reports. One of these concerned a case of escape to England by men who intended to cross in a boat from Monster, south of The Hague. Since links with England were not involved, the report from F2087 was passed to the Naval Abwehr, and the subsequent counter-action resulted in the arrest of three men, among them a certain Gerbrandy, believed to be a nephew of the Dutch Prime Minister in London.

The second of these events concerned a courier line for

agents and material which ran from The Hague to Brussels and Paris. This report required more detailed investigation, for which purpose the IIIF section under Major Wiesekötter used a trick which was employed later in several important cases.

Briefly described, this process consisted of arranging for one of our Abwehr men to be taken on by the enemy group concerned, a man whose experience and knowledge of languages, etc., enabled him to pass in Holland as a refugee Frenchman or in Paris as a refugee Dutchman. F2087 succeeded admirably in recommending our man, Unteroffizier "Arno", to The Hague group as a refugee from the French Resistance, with the result that the group did not hesitate to accommodate him for four weeks in their Hague headquarters. The links and lines which he made in this headquarters were used in subsequent years by "Arno" as controlled enemy courier lines to Paris, Spain and Switzerland.

As I have said, these were affairs not directly touching our main object, but which helped to train my already improving staff.

In the first half of January it was established beyond doubt that the agent group at The Hague must be operating a radio-link with London. The first indication came once more from the reports of F2087. The leader of the group, whose name was still unknown to us, had, it seemed, organised an espionage network whose reports were being passed back by radio. If this were true, the defective transmitter must now be once more in working order.

This was the signal for Heinrichs and his men to redouble their efforts to intercept the traffic. If we desired later on to play this station back, it was essential to intercept every message accurately, in order that we should, in case of a capture, have sufficient material ready at hand so as not to be entirely dependent on the doubtful readiness of the operator himself to collaborate with us. The transmission routine had to be worked out, as well as any planned change of frequency which might be in operation. A number of messages were required, furthermore, in order to establish the nature of the cyphering method, as the plain-language texts would supply us with valuable information required by us for the play-back. From the increasingly precise and more detailed reports of F2087 the aims and extent of operation of this group became clearer every day, and by the end of January it had already become feasible to take action against it, although I was well aware that this could not have affected more than the simple elimination of the group.

A few days after the corresponding report from F2087, the FuB station of the ORPO in Scheveningen discovered that the radio-link with London was in operation via a new Afu (agent-radio) using call-sign RLS. This was in communication with another station PTX. Bearings indicated a position for RLS in south Holland, and close-range D/F narrowed this down to The Hague by the middle of February. There was no longer any doubt that the link RLS-PTX related to the group working under the direction of Captain v.d. Berg, apart from which the station PTX had been "fixed" definitely in a position north of London. This was the English controlling-station for numerous secret radio-links with the Continent.

Continuous and accurate interception watch soon revealed the transmission routine of RLS and the frequencies used, while close-range D/F pin-pointed its position in the south-west part of The Hague. This agreed with reports from F2087 that the set was being operated from a house in the Swellings-traat. F2087 had wormed his way into the position of right-hand man to v.d. Berg, who employed him gladly on all "dangerous" occasions. These included the collection of material for espionage reports, and such enemy questionnaires were provided with answers by us after consultation with the appropriate German land, sea or air authority. Among other tasks, F2087 was commissioned by v.d. Berg to confirm that the cruiser *Prinz Eugen* was under repair at Schiedam. F2087 reported that this was so. The handing of this report to the radio section and its transmission to London were later to play an important role.

According to reports received from F2087 during February, London was pressing for the early preparation of a dropping-area for sabotage material, and it was forecast that the first operation of this kind would take place in the moon-period subsequent to 20th February. We were kept continuously and well informed about the progress of these preparations, and discovered innumerable details, which proved invaluable later on. But all this knowledge would still have been insufficient to reveal the full extent of London's instructions to the sabotage group leader, not to speak of the agreed arrangements prior to the operation itself, had we made a premature move against the group. We had as yet no idea what considerations affected London's choice of a dropping-site, whether moon or dark-moon periods would be selected, how the signal-lights were to be displayed on the ground for the benefit of the aircraft, nor whether white or coloured lamps were to be used. We did now hear, however, for the first time about "positive" and "negative" numbers, which, when they were passed over the BBC's Radio Orange, indicated either times of arrival or postponement of the dropping operation for the current night.

My plan allowed for the first drop from England for v.d. Berg's group to be carried out without interference. But, nevertheless, if we wished to avoid the dangerous possibility that the material might fall into the canals and be lost, a certain degree of invisible and unnoticeable control must be exercised over the enemy. We must in any case be prepared to secure the whole area against surprise.

In February the long, hard winter of 1941-2 had not yet ended. Continuous hard frost had covered all the Dutch lakes and canals with ice, and a thin coating of snow lay over the countryside. Every evening my men and I sat on in the office for an hour or two after work was over to rehearse the parts we should play in the coming operation.

On the evening of 25th February Willy brought a report from F2087 which stated that the nights between 26th and 28th February had been decided for the drop. I took the opportunity to run once more over the position both from our own and the enemy's point of view with my men, including Leutnant Heinrichs:

"We know that a secret organisation from London, consisting of two men, is operating in Holland. No. 1 is the leader and instructor of the group, and No. 2 the radio operator. The names of these men are unknown. We know, none the less, that No. 1 lives in or near Arnhem. No. 2 has quarters in the Swellingstraat which we have identified. It is probable that he transmits from another place situated only a few minutes away. The group is served by the Dutch Reserve Captain v.d. Berg, whose home is well known to us. Through shadowing him we have the descriptions of seven persons, among whom Nos. 1 and 2 must figure. I have circulated all these descriptions to you gentlemen in order that you may be able to impress all details of their appearance, clothing and any special marks on your memory.

"Agent No. 1 has fixed with London for the first drop to be made on an area of moorland near Hooghalen a few kilometres to the eastward of the Beilen–Assen main road. As you are aware, the drop is expected to take place on one of the coming three nights. The executive signal for the operation will be transmitted from London by the use of 'positive' or 'negative' numbers passed over Radio Orange. Leutnant Heinrichs is requested to keep special watch on Radio Orange from now on. When the signal is received, our agent F2087 will at once be informed by v.d. Berg, in order that he can stand by to pick up the dropped material at Hooghalen in a car the following morning. We have made the arrangements for this car. F2087 has been given the name of a policeman in Hooghalen who will be a member of the reception committee for the drop, and he has orders to report to him, using a certain password. When he has collected it, F2087 will transfer the material to a house rented by him in Arnhem. He has my instructions that he is in no circumstances to hand any of this material over to a third party.

"It has been arranged with the Wehrmacht Control Service that the roads from Hooghalen via Beilen, Meppel, Zwolle, Apeldoorn and Arnhem will be exempted from traffic control for twenty-four hours after the promulgation of an agreed codeword.

"Unteroffizier Huntemann will proceed this evening by car to the Beilen–Assen area. He is to await receipt of the code word for the operation and then to inform the local commander at Assen verbally. He is to intervene and to inform me at once should the drop be observed locally, and should anything then be done which might interfere with our plans.

"To avoid any possibility of insecurity, gossip or treachery, I have as yet informed no one about the expected drop and the measures which we intend to take concerning it. After the

58

drop has taken place it may prove impossible to postpone much longer action against the agent group and against v.d. Berg. This action will have in any event to take place if agent No. 1 insists on F2087 handing over the material which he has picked up. Should this happen, it is absolutely essential that the radio operator should be taken by surprise and his transmitter captured intact with all its associated material. Apart from this, it is Willy's duty to identify the operator personally and to discover the probable position of his set. Leutnant Heinrichs is to deploy the three available mobile D/F stations so that the seizure can be effected at the latest during the course of the third transmitting period from today.

"I must emphasise that the Security Police have as yet no knowledge of this undertaking, since it is clear from previous experience that we should not obtain their agreement to our measures and intentions. Should we be forced to arrest the whole group I shall not be able to keep the SIPO any longer in the dark, as the sort of executive action which will then become necessary cannot be taken by us. The SIPO will, however, only be informed at the last moment before the seizure. I am hoping that by this means we shall be able to retain the captured agents in our own hands in order to play the transmitter back with some prospects of success. We are now about to complete the first stage of an undertaking to which you, gentlemen, have devoted all your thoughts and efforts over the past three months. With a little luck there exists a definite possibility that we can continue deceiving the enemy Secret Services in London, even though it be for a short period, by running this agent group ourselves."

Oberst Hofwald was naturally informed about all this. He was very sceptical about the outcome of my plans, but guaranteed his full co-operation and all necessary support against other German authorities should we be forced to take measures which would be incomprehensible to the uninitiated.

All the same, there was plenty of room for things to go wrong. For instance, should the drop be observed by a German patrol it might well result in an armed search of the whole Hooghalen area the same night. Such a mishap would be the signal for v.d. Berg and the agents at The Hague to go underground at once and lie doggo. I had positioned Huntemann in Assen so as to intervene should any suspicious report be received or any question of armed search arise. But he was not to intervene until this actually happened. We had a well-grounded dislike of talking to outside organisations. In spite of all the strict secrecy regulations there was just as much irresponsible gossip in German official circles as there was within the Dutch underground. The desire to make oneself appear interesting and important through having knowledge of secret plans and other matters had brought dire catastrophe to both sides. . . . We often laughed, long afterwards, over our anxiety not to have our plans spoiled by the watchfulness of other German authorities. When the drops came, in the years following, by "conveyor belt", we had ample opportunity to marvel how local German headquarters, sentries and patrols scarcely ever found cause to notice anything out of the ordinary, least of all to do anything about it. After the long day's burden of fat Occupation existence, lying undisturbed by the enemy behind the sure shield of the German radar chain, flak, and night fighters, they obviously slept the deep sleep of the just.

On Thursday, 27th February, Heinrichs reported that at 1330 Radio Orange had repeated the number 783 three times, slowly and deliberately, without any further comment. Events seemed to be progressing normally.

"Keep a good look out, Heinrichs! We don't know yet whether this number is 'positive' or 'negative'. The important thing is to see whether it is followed up. If nothing happens tonight we can expect another number tomorrow or the day after."

At about 1800 the BBC again passed the number 783. If 783 should be positive, v.d. Berg would inform F2087 by 1900, according to plan, that the operation was "on". Willy was then going to ring me from Arnhem, using the code phrase "visit this evening".

At 2200 Willy rang. "No visit expected," he reported briefly. "In any case, George hasn't had an invitation."

"Thank you. Then we must put it off for twenty-four hours at least. Ring me back if anything further occurs and in any case at 2200 tomorrow."

I also spoke to Huntemann at Assen and told him "No visit", but instructed him none the less to keep watch in the vicinity of the "area of the visit" during the night.

On the afternoon of 28th February the tension eased. "963" came clearly, repeated three times, over the BBC. This one must surely be "positive". Shortly after 1800 the same clear, resonant voice repeated over the ether: "Attention—963—963—963." It was like a blast from the horn of an invisible huntsman, giving the signal for the hunt to commence. . . .

We looked at each other in silence, Wurr and I, thinking the same thoughts. This voice had travelled right round the world, and millions upon millions of men must have heard it. But the signal was for two men only. If more than two should understand it, gossiping comrades, unreliable collaborators or we—the enemy—then this message would mean the end of a mortally dangerous undertaking by two brave men. The coming night would bring the first decision in this noiseless struggle, the duel of two adversaries disguised behind their screen. All the signs went to show that our appreciation of the situation was a correct one, and that the reins were firmly in our hands, enabling us to exercise complete control over the first parachute dropping-operation in the Netherlands.

At 2000 the bell rang on my private line.

"Doctor here."

"Willy speaking. I have just had word from George. The captain sent his assistant to him at about 1500 with the news that we can expect a visit this evening. George has to be ready with the car tomorrow afternoon to drive with the assistant to the area. The assistant will drive with him both ways, and he has to set out on the return journey early the day after tomorrow."

"One moment, Willy——" I had to think quickly. This was not in accordance with the arrangements. But matters must now take their course. Any interference would ruin everything, and I had Huntemann well placed to watch the course of events. "All right then Willy! George must carry out the captain's latest instructions. You remain at Arnhem and report back later. Anything more?"

"The assistant arrived in company with a tall, thin man who waited in front of George's house and left again with the assistant. I am certain that the tall man is our No. 1, who has organised the visit for this evening."

"Very good, but let's wait all the same. Good luck!"

I soon got Huntemann on the line and told him about developments. "Keep a careful look out, because George will not be with the party this evening. But don't take any risks, and let me know tomorrow morning whether the visit takes place or not."

That evening, 28th February 1942, the first drop of weapons and sabotage material took place in Holland, eastwards of Hooghalen, as agreed and planned with London. The aircraft, a twin-engined bomber, circled over the dropping-point at a height of scarcely 600 feet, dropped two containers on parachutes, and disappeared again westwards. I heard this from Huntemann in the morning. No German posts had heard anything suspicious, and the local headquarters at Assen knew nothing. . . .

Everything else went smoothly. F2087, accompanied by the assistant, whose name was Biermann, collected the dropped material in his car from the vicinity of the dropping-point, and by the next day, Sunday, it was all "safely" in his house. The removal of the patrol from the main road had fully justified itself.

Willy arrived in the evening. "All well, Chief," he reported. "But one thing has gone wrong. The Reception Committee has only picked up one container, as the other was carried away by the wind, although they have searched all night without result."

I stared at him in horror. What on earth could have happened? According to Biermann's account which he gave to George, at the dropping-point when the container came down, apart from himself and the captain who was directing the reception arrangements, there had also been "tall Thijs", as he called his previous day's companion. Thijs had placed three men in the form of a right-angled triangle, each with a powerful red pocket torch, the long side measuring 120 paces in length. He had stood himself at one corner with a powerful white torch and gave orders for the four lights to be shone in the direction of the aircraft as it approached. At about midnight an aircraft came in flying low, turned towards the light signals and flew over the triangle, releasing a container as it did so, which landed safely close by.

The aircraft immediately gained height and released a second container, but this had been carried by the wind towards the Beilen–Assen road and could not be found even after hours of searching. The single container which was picked up contained explosives and fuses of many kinds, also Colt automatics with ammunition, of which Thijs took two. The remainder, together with the parachute, was buried on the spot for the time being and brought to Arnhem by George the next day.

Willy's report was interesting and full of new possibilities, but what had happened to the second container? If it fell into the wrong hands we should be in a mess! Although bad enough if it fell into the hands of the underground, if it were picked up by the German authorities it would put the "sleeping dogs" in the whole of Holland on their guard! Eventually we hit on a good idea. There was quartered in Assen a naval-training battalion about a thousand strong. A friend of mine, Oberleutnant von Boland, who was serving as Luftwaffe Abwehr officer in Holland, and whom I took into my confidence, paid a visit to the Battalion Commander and spun him a yarn about a mishap, involving secret matters, which had occurred in the Luftwaffe. A valuable new piece of apparatus had fallen from an aircraft during a night test flight and should have landed somewhere to the eastward of Hooghalen. It was of vital importance to find it. The Commanding Officer willingly arranged to carry out a two-day field exercise with his battalion in the Hooghalen area, and a reward of five hundred guilders was offered to the group which should find the missing article. The exercise duly took place on 2nd and 3rd March, but unfortunately without results. Finally we consoled ourselves with the thought that the lost container must have fallen into a dyke or a canal, where it could safely be left. In any case we never heard anything more about it.

After the drop the radio traffic from Afu RLS increased considerably. Several messages passed nearly every day to London with an equal number from "over there", so that Heinrichs' D/F detachment had plenty of opportunity to work on it. On 4th March Heinrichs reported to me that RLS could be pinpointed with reasonable certainty in a modern block of flats in the Fahrenheitstraat at The Hague. The time for a seizure had arrived.

We knew more about this particular Afu group than at any time during previous actions against secret transmitters. The next routine transmission would be on Friday the 6th at 1800 hours, and I laid my plans for the seizure to take place at this time.

In Arnhem conditions were not quite the same as previously. In the days succeeding the drop "tall Thijs" had been seen on several occasions near George's flat—he had probably arranged to have him watched. Did this mean that he did not altogether trust George? Or was it simply to make sure that the material was not removed without his knowledge? The captain had told George that the sabotage material would not be left at his house for more than a week. Steps must be taken to deal with it before then.

I arranged a meeting with Schreieder for the morning of 6th March to discuss a forthcoming major operation and certain questions of principle touching it. I had had all the results from operation *Nordpol* up to date, together with details of personalities involved, etc., set out in a single report for the use of the SIPO. Results obtained from the intercepted radio traffic were not included, as I had decided to lay down a clear line of demarcation which should govern any future concerted action between us. Arrest, interrogation and the further investigation of leads would remain, as hitherto, the province of the SIPO. Similarly the planned continuation of the *Nordpol* operation, and in particular the playing back of RLS, must remain the exclusive responsibility of the Military Abwehr, that is to say of my own section, through IIIF.

Schreieder received me with his particular, eel-like, ingratiating manner in the bright conference-room of the Binnenhof. His immediate superior, Sturmbannführer Wolf, was in attendance, probably on account of the "certain questions of principle". Wolf was a typical product of the SS, a man barely thirty years old, who had worked his way up in a short time from solicitor's clerk to SIPO Regierungsrat, and who made up in arrogance for the specialist knowledge which Schreieder had gained from many years in the Criminal Police. I found his utter mistrust of the Wehrmacht, and particularly of the Military Abwehr service, almost stimulating, and his adolescent conceit a constant source of quiet amusement.

We lit the cigars which Schreieder passed around, and I began as follows: "I have set out in a detailed memorandum the course of the events which have led up to the action for which I am now asking your support. The ramifications of this matter are rather wide, and I would ask you to glance at this paper before we proceed any further," and I handed to each a copy of the memorandum.

They started to read, curiously and coolly at first, but soon began to turn the pages quickly and excitedly. Wolf raised his eyes first and went straight to the point.

"If your F2087 had not actually got the material at his house, as you state, I should not have had much faith in the rest of this," he said. "In my opinion, firm action should have been taken long ago against v.d. Berg's activities. If any use is made of this material which might endanger the internal security of Holland the responsibility is yours; and may I draw your attention, Herr Major, to the fact that everything in the interior which concerns illegal activity is the responsibility of the Sicherheitspolizei? In particular, you should long before this have reported the gang which went to meet the aircraft."

He leaned back in his chair and attempted to register disapproval. Schreieder stared in front of him. His cigar evidently claimed more of his attention than the pronouncements of his youthful superior whom he was not disposed to take very seriously.

"My dear Herr Wolf," I replied, "I obviously accept *in toto* your objections of principle against the course of our actions in respect of *Nordpol*, but I would ask you to remember that as in love, so in war, nothing succeeds like success. In either case, once success has been achieved the means used and the rights and wrongs of procedure, looked at retrospectively, have little importance. The important question is now whether your act of seizure can be carried out in such a way that no loose ends will be left which might endanger our subsequent play-back. I leave this problem, gentlemen, with confidence in your experienced hands. As soon as the arrests are made I will withdraw my men from their present activities and leave you a clear field. I hope soon to have other work for them to do."

Schreieder rose and took from a cupboard town plans of Arnhem and The Hague. The brain of the criminologist was at work, and, his pad soon full of notes, he exposed his plans concisely and clearly: "Friday at 1800, capture of Afu RLS. A.m. Saturday, arrest of v.d. Berg's group. Simultaneous action in Arnhem to secure Agent No. 1, Thijs." For the time being no action was contemplated against the Reception Committee in Hooghalen. The flats of the arrested men would, however, remain occupied, and anyone who should enter them would be arrested. All was quite clear up to this point. On the second point, namely clarification of responsibilities for the play-back of Afu RLS by Abwehr IIIF, Schreieder voiced no objections, and Wolf as well was noticeably conciliatory. After their unfortunate experiences with the operator Zomer and v.d. Reyden they were not disposed to grudge the Abwehr an opportunity of burning its fingers by attempting the play-back of a transmitter.

I pocketed the receipt which had been signed by Schreieder for the Top Secret memorandum handed to the SIPO, and after assuring them of my readiness to discuss the paper further with them at any time I left the building.

An icy north-west wind was blowing beneath low, leaden clouds around the street corners of The Hague, and the cold dampness of the air was intensified by the coming of darkness. I walked slowly westwards down the Fahrenheitstraat, which lay before me broad and deserted, with only a few pedestrians to be seen hastening homewards. In the last block of houses in the street, just before it reached the edge of the green belt, stood the building containing the flat in which Station RLS had been pinpointed by D/F bearings. I wanted to look over the area of operations once again before the seizure. Six or seven broad stone steps led into the deep porches up to the doors of the flats. At least six such doors led into the block, but behind which was the radio set?

According to Heinrichs, the suspected flat was somewhere in the centre of the block, and I could scarcely believe my eyes when I made out a familiar figure in one of the entrances. Willy! Exhausted and frozen almost stiff with cold, he was leaning against the wall. "Chief, he's in there—the operator, I mean! It must be him. I was watching the suspected house in the Swellingstraat this afternoon, when just before four o'clock a man came out who fitted fairly well the description we have of him. Middle twenties, fair, medium height, small face, spectacles, raincoat. I followed him this way, and where does the fellow go but straight into this house here, the probable transmitting-station! He's been inside for nearly two hours now, probably cyphering and getting his set ready for the next routine."

"You've done the trick again, my friend," I said, looking at my watch, "it's 1750 now. In ten minutes we're off, and don't fall down until then! I must get to Heinrichs, who should be just round the corner with his D/F car. As soon as this man gets on the key and tries to establish communication with 'over there' we shall strike."

At the western end of the Fahrenheitstraat, concealed by the corner building from the suspected block, there stood a large car with Dutch number-plates. In it there were two men in plain clothes, and a third—Heinrichs—stood in a gateway a few yards away. A curious passer-by would have had to approach quite close to the car and to look through the window in order to see that one of the occupants was busy with some apparatus on the rear seat which involved the slow turning of a knob.

I entered the gateway and told Heinrichs shortly what Willy had reported. "I'd like to have a bit of luck like that," he laughed, "but the main thing is that I now know that we have 'fixed' him correctly—and we can get right at him once he touches the key!"

One of his men motioned Heinrichs to the car, and as he lowered the window a short way a soft piping note became audible, repeated at brief intervals. The key-clicks of a transmitter working very close at hand were breaking through, although the receiver had not yet been tuned to the exact frequency. "RLS working and trying to make contact," remarked someone from inside the car, and the window was closed again.

I looked at my watch. 1801. At 1800 the SIPO should have been in position in the next street waiting to hear from me that the transmitter was operating. Now was the moment to act·

quickly and noiselessly before communication could be established with London, in order to prevent any warning indication from being given by the operator in the nick of time about the seizure. "In two minutes' time by the front-door!" I called to Heinrichs, and hurried round the corner into the deserted Fahrenheitstraat to the doorway where I had found Willy.

But no Willy! Where could he have got to? I broke into a run, meaning to fetch the SIPO, when I suddenly noticed Willy, a block away, running eastwards along the Fahren-heitstraat. I sometimes used a signal which all my men knew. When they heard this signal, which varied between the croak of a raven and the cry of a seagull, everyone knew that I was nearby. It echoed along the quiet street. Willy stopped, turned round, and, seeing me, raised his clenched fist three times vertically in the air—the signal for full speed. I arrived at the run at a corner where a gang of SIPO men in plain clothes were in the act of jumping from a row of cars.

"Quick—the first car this way!" I called, and signalled the driver over to me. Three policemen jumped in as the car moved, and I jammed myself in beside the driver. "Over there, to the man in the raincoat!" and I pointed towards Willy, who was hurrying on down the Fahrenheitstraat. Before we got up to him he turned and pointed at a figure that had crossed the road a hundred yards ahead of us and was disappear-ing down a side street. "After him!" Round the corner and there he was, hurrying onwards without looking behind him. As we roared up he stopped dead in his tracks and turned with his back to the wall, his pale face turned towards the four men who sprang from the car and surrounded him.

"German Security Police! You're under arrest!" Two of the men had drawn their revolvers while the third searched him quickly for arms. It was the young fair man whom Willy had described to me, the man who had gone to the Fahren-heitstraat at 1600. "Get in!" It was a seven-seater car and they took him in behind and between them. I again sat beside the driver, and we drove back to the suspected house. The whole affair had lasted scarcely a minute, and the Fahrenheitstraat was again empty.

After the precious bird had flown—the bird which so soon

afterwards we were able to catch through Willy's presence of mind—the SIPO, led by Heinrichs, had forced their way into the house and carried out the main object of the raid. A man keeping a look-out had informed the operator of the arrival in the side street at 1800 of four large cars with six men in each. The watcher was his friend Teller, the owner of the flat in which the set was situated. The operator did the one thing possible by leaving everything where it was, grabbing his coat and dashing out of the house. He could have scarcely known that a man who knew him was outside, who would shadow him.

Teller and his wife, who were both arrested, had quickly thrown the set and all the papers into a small trunk, which they had dropped out of the window into the back garden. It had been found there, quite undamaged and resting peacefully across two lines of washing!

Heinrichs went carefully through the captured booty. "Everything's here, Herr Major. Messages, a lot of cypher material and the crystals. We'll get on to questioning the agents at once and have it all cleared up. Above all we must decide one thing. His next routine with London is on Sunday at 1400. If he is wise, he'll go on the key himself, but if not I will have one of my own operators who has listened to him transmitting so often that he can imitate his 'handwriting' on the key without much danger of detection."

Heinrichs was in his element, no doubt foreseeing a great chance for the little lieutenant of the ORPO radio-interception service to prove once and for all what a genius for radio and cypher work was concealed in him, and that ten long years' efforts on this special territory had not been in vain.

During the previous few days we had decided to go ahead with the play-back, provided there was sufficient captured material, and with or without the services of the agent-operator. I had entrusted the care of the arrested man, whose name was Lauwers, to Willy in separate quarters, and I stayed a short while to listen-in to the interrogation and to get an impression of this agent. Willy, who knew my intentions perfectly, was endeavouring with some skill to use the understandable shock sustained by the arrested man to establish a firm basis of confidence. It was important not to treat the prisoner with the severity, hostility and crude

69

methods which were the nightmare of such agents if caught by the Germans, but a friendly understanding should be shown of his proved operational ability, and human sympathy for his misfortune. Willy didn't have to act a part such as this. It came naturally to him, a fact which in itself reduced the danger of failure.

Lauwers had meanwhile regained his self-control. With the acute perception of a hunted animal he doubtless detected the special nature of our attitude towards him, which did not in any way fit his preconceived ideas of such a situation. I reassured him about his personal position and was soon able to confirm that he spoke excellent German. He asked that the arrested Teller family should be spared severe treatment, as it was he who had persuaded them to provide the cover for his radio activities. I sent Willy with him back to his isolation cell in Scheveningen prison, and set myself to prepare for the probably decisive interview with this man which must soon take place.

Schreieder co-operated sensibly, and all my proposals were well received. He and his SIPO comrades were now convinced that the operation would have been out of the question but for the preparations made by IIIF, and he and Heinrichs kept me well informed about the interrogation of Lauwers, which made rapid progress. Apparently the latter was talking quite freely. He had been in business in Singapore at the outbreak of war, had come as a volunteer to England and had been recruited by a certain Captain Derksema into the Anglo-Dutch Secret Service. After courses of instruction in radio and parachute work he had apparently been landed on the Dutch coast, together with another agent, from an MTB. As his entire previous career as an agent could be easily established, he wisely made no attempt at telling lies, and his questioners, who included Willy, came eventually to the conclusion that his statements were reliable. The man found little to quarrel about, for after his interrogation by Leutnant Heinrichs it soon became clear to him that all details of his operating and the particulars of his transmitter were as familiar to us as they were to himself. We hoped soon to unravel such points as were not yet clear.

On the same evening as the seizure the SIPO had taken

action against Captain v.d. Berg's group and had arrested
them and all their known collaborators. Thijs was in Arnhem,
and his arrest was expected hourly.

The second Sunday in March seemed like a herald of
spring after the cold wet weather of the previous weeks.
White clouds sailed across a pale-blue sky, and the March
sun exposed mercilessly the damage caused by the long frost
and winter storms in the garden of the Abwehr mess. I was
sitting after lunch sunning myself in a sheltered corner when
a large, powerfully built man came out, his open raincoat and
fair hair blowing loose in the wind. He could only have been
looking for me, for he knew no one else at headquarters.
At my wave and call of "Hullo—Hermann!" he drew off
his heavy driving gauntlets and crossed the lawn in the direction of my corner. Our hearty greeting, as between two old
friends, was soon cut short when he asked if he could discuss
with me a matter of some urgency. He had arrived at The
Hague at midday and apologised for not having reported
himself, but there was a special reason for his hurrying on to
see me at Scheveningen. His customary quiet self-confidence
was tinged with a noticeable disquiet, and as he made this
request, his blue eyes wandered uncomfortably. Evidently
something serious was up.

"Let's go to my room," I said, taking his arm. "What
have you got on your mind?"

The question with which he countered mine halted me in
my tracks with astonishment. Had I been struck by lightning
I could not have been more surprised. "Do you know a man
called Lauwers?"

I walked on silently ahead towards the back entrance of
the office, with thoughts whirling in my head. What could
have gone wrong?

This old friend, who was some years younger than myself,
had had a very successful rising career as a merchant and
industrialist. Since 1940 a German ministry had used his
specialist and language qualifications to rehabilitate the can-
ning factories in the Western Occupied areas. He had led
since then an unsettled kind of existence between Berlin, Paris
and The Hague. He had a superficial knowledge of the type

of work in which I was engaged, and shortly after my transfer
to The Hague had paid me a social call.

"How did you come across the name Lauwers?" I asked
as we sat down.

The story I now heard was the result of an improbable chain
of pure accidents such as one only meets in real life, for no
film producer would have dared to invent and screen such
a chain of events. On the morning after his arrival at The
Hague Hermann had telephoned a lady whom he had known
well for years as the secretary of his best Dutch business friend
to ask her to lunch with him for a discussion of urgent business
affairs. Instead of this, she asked him to come and see her
at once, and when he did so he learnt that she expected to
be arrested at any moment by the German Security Police.

Some months previously she had given shelter to a young
Dutchman who had arrived from England to work as a secret
radio operator. This man had changed his quarters several
times since then for security reasons, but she had heard only
this morning that he had been arrested. His name was Lauwers.

"Do you know this name?" repeated Hermann. "And,
anyway, can anything be done for this lady?"

I sat as if someone had hit me over the head. Here were
all the right ingredients for pure melodrama: the brave
spy with his radio set working behind our lines in disregard of
death; the romantically inclined young woman, more influ-
enced by the clever spy than by the death penalty which could
be exacted for sheltering enemy agents; the faithful friend with
contacts in high circles, who appears in the nick of time to
prevent the worst from happening; and, finally, the enemy
Abwehr officer, who, against all considerations of duty and
conscience, will save the romantically inclined young woman
from the firing-squad! Perfect!

How could it have come about that I should again have to
play the ludicrous role of guardian angel for grown-up children
who wouldn't listen? It was true there was reason to hope some
grass might have grown over this "Operation", but any stray
cow might at any time come and nibble it away again.

I brooded over the problem for a longish while.

If I were to save this woman from court-martial or from
years of disappearance into prison a great deal would depend

on her future conduct. If the SIPO were to get a glimpse of the truth nothing could save her. Lauwers would certainly hold his tongue, but who else was there who knew of this connection? Everything hung on whether she was clever enough and had sufficient self-control to forget that she had ever known an agent named Lauwers now under arrest. With horror I suddenly remembered our plans for the play-back of Lauwers' transmitter. She could scarcely know our intentions, but an accidental remark from her could ruin everything. And if the fact then became public that I had known of her connection with Lauwers the consequences were not pleasant to contemplate.

Hermann had walked to the window and was gazing out—my long silence had become hard for him to bear. The sun's rays were striking palely across the bare treetops. What, in fact, must I do? Mustn't I have her arrested forthwith, like everyone else who was aware of the name and the fate of the man Lauwers?

My voice, as it spoke, sounded strange in my own ears: "I can do nothing for this lady. Only she can do anything and that is *keep silent*! I rely on you when I assume that she will not intentionally pass on her knowledge. Should she, however, breathe a word to a single living soul then nothing can save her. In such an event she will not see daylight again until the end of the war, provided that nothing worse befalls her. Go and tell her that, and please do not say to whom you have been speaking. I advise her to go away at once for a fortnight 'in the country'. Hurry there and tell her, before it is too late."

I accompanied Hermann back through the mess garden. "You may have my word all will be well," he promised me, as his big car shot off down the street.

The rest of the afternoon gave me an opportunity to consider whether it was possible for an Abwehr officer to behave decently, as the saying goes, or whether he was an idiot to do so. The future would decide. In any case I was pleased to have again been able to resist the temptation of playing the public prosecutor. (I can assure the anxious reader in advance that the lady kept her freedom, since she knew how to hold her tongue. What's more, she used this freedom to give aid and comfort to her fellow-countrymen who had gone underground. But *that* I didn't know until after the war.)

Chapter II: The Drama

A high and slightly-curved brick wall runs for many hundreds of yards down the Alkemadelaan in Scheveningen and separates the extensive solitary confinement prison of The Hague from the outside world. Shortly before noon on the third Sunday in March two men emerged from a small door beside the massive iron gates which alone break the continuity of this wall. Jumping into a small closed car, they drove off down the broad sunlit surface of the quiet street. Skirting the residential quarter of Scheveningen, it eventually stopped in front of a large detached house in the Parkstraat, the headquarters of the Radio Interception Service of the ORPO.

It was my first drive with Lauwers, whom I had brought with me, after a long personal discussion, to the FuB headquarters. The time had nearly arrived for the fourth routine transmission by Afu RLS since the seizure of the set, a transmission which must at all costs bring about our first radio contact with London by the set under our control. Since the seizure we had let two routine periods pass without using the transmitter. When London called RLS on the third occasion, one of Heinrichs' expert operators had gone on the key for a short period, but had then cut short the routine making the excuse of "atmospheric interference". Willy had reported a few days previously that, in conversation with Lauwers, he had received the impression that the latter might be induced to co-operate in playing RLS if approached by a responsible authority.

This forenoon I had accordingly sat opposite him for two hours in the small visiting-room of the prison.

After a few introductory enquiries touching his health and living conditions, I put to him forcefully that he alone could assist me in my plan to save him and Thijs, who had meanwhile been arrested, from sentence by a German military court. I used every kind of appeal to his intelligence and emphasised how pointless it was for him to throw his life away. He must at least help me to produce one single argument which would enable me to avert the death sentence that would otherwise certainly be passed on both Thijs and himself. To do this he would simply have to transmit today at noon the three messages which he had been unable to pass on 6th March when he had been arrested by us.

Hitherto Lauwers had spoken scarcely at all, but now he showed some interest.

"What? I must pass my last three signals?"

"Of course. You can't object to that, and you can do it with a good conscience into the bargain. As I have already told you, I can have it done by one of our own men, and much more safely from my point of view. But in that case there will be no option for you and Thijs."

"Yes, but do you know what is covered by those messages? They might have very dangerous information in them from the German point of view."

"I know just what London can be told, and what not. These signals are of no great importance. For example, there is no harm in their knowing 'over there' that the *Prinz Eugen* is lying at Schiedam. Incidentally, how did you come by that information?"

Lauwers pursed his lips and looked vacantly past me through the high, barred window. He sat silent for a while. In these eight days spent in prison he had grown thinner and paler. Anxiety for his associates, for his friend Thijs and for his own unknown destiny had left its mark. I didn't break his train of thought, for I felt that something was about to emerge. Perhaps he was afraid, by answering my questions frankly, to betray a line which had so far remained concealed. However, his answer was not very important, for this report had in any case originated from our own sources. And Heinrichs had already deciphered so many of the captured signals that we knew the general nature of their contents.

"Let's leave it," I said. "In any case it doesn't matter a great deal. Have a cigarette."

I pushed across my cigarette-case. I had noticed at the time of the arrest that his fingers were stained with tobacco, and now he drew the smoke deeply and hungrily into his lungs, his thoughts evidently less far away than they had been.

I changed the subject. "You see, I am a soldier and no

politician," I said. "Nor am I an investigating judge. The penalties for your existence and activities as an agent don't interest me. As far as I know you are a Dutchman with the English rank of lieutenant. I am treating you as an officer despite your plain clothes and in spite of the circumstances in which we have met. I am sorry that in your individual case the regulations do not permit me to accord you the treatment customary in the case of officers. As a soldier, I have the greatest respect for your courage and devotion to duty, but I must say none the less that the kind of job which you have been given by London is not very respectable.

"However, I admit you are a soldier, and so must obey orders. It is fully understandable that London should try to organise an Intelligence Service in the Netherlands, in which the only people who risk their necks are those directly connected with it over here. But a mode of action involving the arming of civilians and the use of explosives, that is to say the starting of a war of *franc-tireurs* and shooters in the back, is a different matter. I don't know where, in what colonial or Balkan set-up, the men who have thought out this new form of warfare for Western Europe can have obtained their experience. In a country as civilised and as thickly populated as Holland I personally consider such methods to be highly dubious. The possible advantages of this kind of warfare can never bear comparison with the consequences of reprisals by the Occupation authorities as justified by the laws of war. Any Army of Occupation, irrespective of its nationality, will crush such attempts by means of the shooting of hostages and the terrorising of the population. I have therefore decided to prevent the Allied Secret Service, by the use of all the means at my disposal, from supplying tons of arms and explosives to irresponsible fanatics in this country, the use of which can only mean a blood-bath for the civil population."

Lauwers had listened to me doubtfully and with growing surprise. Did he think that my frank indignation was not genuine? I paused to give him time to answer, but as nothing happened I went on.

"You probably understand better now why I am venturing to ask for your help. I don't know yet whether I can carry my intentions into actual practice. No doubt it will soon come to

76

London's attention that the German Abwehr is sitting behind the key of RLS, and then we shall have to start all over again. In any event they will have been taught a lesson which may delay their plans considerably. Neither you nor I know when this senseless mangling of one side by the other in Europe will come to an end, but I am quite certain that no 'victor' is going to emerge from it all. All that will be left will be the crippled, the blind, the lame and the one-armed. Neither side will be able to survive in the future without the support of the other. Let us try to wage this war so that we can look one another in the eyes at the end of it. Anyone who helps me in this I will count as my friend, even though he may wear a different uniform."

The clock pointed to a few minutes before twelve. The time had flown. I rose and put on my overcoat.

"We must get ready for the routine transmission at 1400. Are you coming along?"

Lauwers had also risen and his eyes sought mine. He replied loudly and clearly:

"Yes."

And we went out together.

At the FuB station I handed Lauwers over to Heinrichs and his men, who made the new assistant welcome and were soon exchanging views in coding and radio-technical matters. Just before 1400 Lauwers put on his earphones and sat down before his old set, just as though there had been no interruption in his work. Heinrichs, of course, had taken his precautions against the possibility that Lauwers would betray us through some quick signal. One of his men close by was listening-in to Lauwers' transmission—his hand ready on the key of a jamming transmitter.

Our first routine with London went off normally. We cleared the three captured messages and then ourselves received several signals which referred to previous reports from RLS. Nothing unusual occurred—there were clearly no doubts in their minds "over there". The radio-link between the German Abwehr and the Allied Secret Service in London had been established. But how long could it be maintained? It was by no means certain we wouldn't be found out after the next routine period. The critical factor was the cyphering. If there should be signs in it which, unknown to us, were to confirm the genuineness of the operator—so-called "identity checks"—the bubble might well be pricked after our second transmission. In that case it was conceivable that London might try to "play" on its own— and we should never know what mistake or trap we had to thank for our being discovered. While decyphering some captured and previously intercepted messages from Lauwers Heinrichs had noticed that every third *stop* in the English version was written as *stip*. Lauwers, however, declared this to be his own identity check, and we continued to use it subsequently over RLS without protest from "over there".

I returned to the mess in excellent mood and reported the first developments, but Hofwald was sceptical as ever.

"I don't want to belittle your efforts," he said; "the capture of the set and the sabotage material are certainly something— but let's wait a little. I don't believe it will go on. London will find out somehow that you have lifted their nest in this country. Above all, take care that no one is playing with *you*. And don't let's be in too much of a hurry to start cheering."

In my heart I fully agreed with him. From the outset we took special precautions to ensure the security of our link. At first I only informed the Abwehr headquarters, OKW Berlin, and Radio-Abwehr headquarters Berlin, which had to be done in any case. No other authorities were told.

The second routine, on Wednesday, brought us a surprise, for in it London requested the early preparation of a dropping zone for a large quantity of material. Furthermore, an agent would be dropped during the forthcoming moon period, about 25th March, and must be met by a Reception Committee from RLS. Code word for the operation—Watercress.

Triumph! If this should come off the genuineness of Operation *Nordpol* would be finally established. We had no previous experience of such operations, and nothing must be left to chance if it were to succeed. We were soon in trouble, however. Lauwers, naturally, had been present when these important messages were being decyphered, and the news that we were to receive another agent upset him considerably. He told Heinrichs firmly that he would have nothing whatever to do with it and that he refused to operate any longer, which, after trying vainly to get Lauwers to change his mind, Heinrichs reported to me.

Two hours later Lauwers repeated his refusal, sitting opposite me in the waiting-room of the prison. I must realise, he said, that he could not be responsible for allowing comrades to fall into our hands. He would co-operate if London were to send across sabotage material, but not if agents were to be dropped. The obstinacy of this little man was remarkable in view of his position—we were evidently in for a long discussion.

"Rest assured," I told him, "that we can lay our hands on this man whether you co-operate on the set or not. You know as well as we do that London will only send him if they have full confidence in the radio-link. And after today it is surely obvious that this is the case. Now listen! I have already proposed to OKW Berlin that you and Thijs should not be brought before a military court, but kept as prisoners until the end of the war. I am confident that this proposal will be approved. If you continue operating for us I hope to obtain a decision from the highest authorities that none of the agents who may fall into our hands through Operation *Nordpol* shall be given the death penalty. Whatever term of imprisonment they may receive, they will be set free at the end of the war, which is the main thing. You can do everything possible for this man if London sends him over. You know it is my duty to defeat London's intentions of starting a *franc-tireur* war. It's already something if we can capture their material, but for that very reason we can't leave any agents at large. Every man who is sent across for such work must be rounded up. The end justifies the means. I believe that it is better to lock up London's agents than to subject the whole population of the Netherlands to a bloody reign of terror. Think over your decision once more. In any case I will see that you are brought to the FuB station at the time of the next routine."

Lauwers remained silent. I had said all that had to be said and had no wish to bring pressure to bear on him—he must make his own decision.

When he had gone I had the guard in, a bear-like SIPO man in the uniform of an Obersturmführer to speak to me, and learnt from him that he had received instructions from Schreieder regarding the decent treatment which was to be given to Lauwers and Taconis, the latter being the correct name of "tall Thijs". Unfortunately, things were going badly as regards Taconis, and the warders were afraid to be alone with him in his cell.

"Please have Taconis brought along here," I said. "I should like to get to know him better."

Two warders led in a tall, thin young man, whose Indonesian blood could scarcely be detected, and remained by the door. He wore handcuffs, and his large, dark eyes regarded me haughtily.

"Sit down," I said, and, turning to the warders, "Why is he handcuffed? Remove them at once. I don't wish to sit opposite a handcuffed man."

"He attacked a warder yesterday and attempted to escape," said one of the warders, "and the Commandant has given orders that he must be handcuffed when out of his cell."

"Do you understand German?" I asked. "Is it true what this man says?"

The prisoner nodded.

"Take the handcuffs off, on my responsibility," I ordered; and as they hastened to do so, "I hope you will be more careful in future," I told Taconis, who seemed indifferent to what lay before him. "If you try to beat up warders you mustn't be surprised at being put in irons. I know that you are an English lieutenant and what your duties have been here in Holland. I respect your courage and understand why you have hitherto refused to talk, so I can give you credit for sufficient self-control not to make unnecessary trouble for yourself and the warders in future. If you wish to speak to me or have any special requests to make, please ask the Commandant to have me informed. And meanwhile I shall be grateful if you are sensible. Understand?"

Taconis again nodded indifferently, a gesture which did not disguise his fiery temperament. When almost past the door he wheeled round suddenly and gave me a slight bow, which I returned with some astonishment. A sporting, tough character. I hoped he had understood me.

I now formed a special group within IIIF headquarters, under my personal control, for the further exploitation of *Nordpol*. Wurr, Willy and Huntemann were included, and at the start Oswald as well. I turned the remainder of my staff

over to Major Wiesekötter, with independent responsibility
for handling leads and contacts, which left my own hands free
for the main operation. This basic form of organisation was
retained throughout the subsequent vast increase of work and
the consequent expansion of the IIIF office. Duties were allo-
cated within the group as follows: Oberleutnant Wurr—
Adjutant and Deputy for Operation *Nordpol*—was in on everything and knew everything, in case for any reason I should
fall out; his duty was to cultivate contact with Section IVE
of the SIPO so as to have knowledge of all points rising from
interrogations. Huntemann was Liaison Officer and adviser
to Heinrichs, with a special responsibility for captured agents;
he assisted me in preparing our messages for London and was
responsible for their correct translation according to the
vocabulary and literary style of the agents concerned. Oswald
acted at first as his assistant, but was soon afterwards transferred back to Major Wiesekötter's section. Willy was ear-
marked for duty in the field and for the further exploitation
of F2087, for whom I had a special plan in mind when we
had received the agent whose impending arrival had been
announced.

Communication over RLS passed off normally during the
routines which followed. The signal traffic increased more and
more, being chiefly concerned with our proposals for the new
dropping area and the part we should play in connection with
the drop itself. I had not seen Lauwers again, but Heinrichs
had reported that, in spite of his previous refusal, he had been
back on the key since Friday the 20th.

We had learnt from signals from RLS intercepted previously
that a special method was used for communicating the exact
position of a dropping point. Besides the Hooghalen area,
Taconis had proposed another dropping point to London close
to the southward of Zoutkamp. This position was passed by
means of a six-figure number, the figures being written and
coded into the text of the message. During the seizure we had
captured the whole of the maps used by the RLS group, among
which was a complete set of ANR maps, with the positions
marked in together with their coded symbols. With the use of
the earlier messages intercepted it was then no longer difficult
to work out the method of coding positions.

We now told London that the position near Zoutkamp was considered to be too isolated, and we proposed a new position, which Wurr and I had chosen, on the moorland about three kilometres north-east of Steenwijk. Near Steenwijk, as it happened, was stationed another naval training battalion, in barracks built just before the war. The Commanding Officer, who turned out to be an old acquaintance, co-operated with us splendidly in the subsequent period, during which numerous drops took place in the Steenwijk area. I took him fully into my confidence from the start and kept him informed of my intentions in case any posts or patrols should report night-flying activity over the moorland near Steenwijk. Should this happen, he was to issue a cover-story to the effect that Luftwaffe night-flying instruction was taking place in the vicinity.

On 25th March London signalled its agreement to the new proposal.

Subsequent signals informed us that the drop could be expected at any time after 27th March and that the actual day would be indicated as previously by "positive" and "negative" numbers. "Abor" would be the name of the agent who was being sent. "Abor" would know that he had been met by "Ebenezer's" men if he was greeted by the use of this name. ("Ebenezer" was the code name for Lauwers and Afu RLS.) The reception party must be prepared to take away four large containers. One of these would be marked with a white cross, and this contained a special parcel for "Ebenezer". These messages looked distinctly promising, and I now had few doubts that things were about to develop nicely.

On 26th March I gave Schreieder a report on the situation, in which I requested SIPO co-operation in the arrest of agents whom I was expecting "from the air". Schreieder's normally expensive-coloured nose grew a thought paler as he read the report. He excused himself for "just a moment" and rushed in to his senior officers, Wolf and Dr. Harster. When he emerged after ten minutes he had regained control over himself—he had clearly received a "bottle" for taking too seriously his duty of telling tales about the Abwehr. I took advantage of his surprise, however, to have our respective duties and responsibilities laid down clearly in advance. Since it was going to be a matter of regular routine for SIPO and Abwehr to work together in the course of such night operations, all possibilities of friction must be ruled out from now on. We agreed that the SIPO alone should deal with the agents themselves, but that the senior Wehrmacht officer on the spot should be in charge of the entire operation in the capacity of Commandant of the whole dropping area. It was not necessary to dwell on the necessity for strict secrecy precautions and the need to limit strictly the numbers of those "in the know". Schreieder was much too skilful and experienced a policeman not to appreciate the great potentialities of a radio-link with the Anglo-Dutch Secret Service in London. From his point of view an operation like this had a number of side issues which had escaped my attention, since I was naturally more concerned over the security and intelligence aspects.

When we received the "positive" number over Radio Orange on the afternoon of 27th March it was no longer much of a surprise. This simply confirmed the solution of our difficult mathematical problem as exactly correct. The equation had come out, and no unknown quantity was left.

At about 1900 I was driving with Wurr in my car, the headlights boring their way through the evening darkness towards Utrecht. I had arranged that we should all meet at 2230 at the house of the Naval Commandant at Steenwijk. In addition to the SIPO I was expecting Leutnant Heinrichs with three of his men. I had promised him that the ORPO should look after the first triangle of red lights to be set up in Germany for the reception, by arrangement, of an English aircraft. Willy also turned up from Arnhem with a passenger, F2087, who was to welcome Abor in the name of Ebenezer. . . .

At about 2300 a small string of cars left Steenwijk with dimmed lights and took the track across the marsh and moor-land in the direction of the dropping area, until they were lost to sight in the thick undergrowth of a small wood close by. It was getting on for midnight. After putting the lights of the triangle ready in position the handful of men was swallowed up in the shadow of the juniper bushes and the general dark-ness and sat discussing the prospects of the operation in under-tones. I had forbidden the use of lights or loud conversation, and what it amounted to was that everyone had to behave as though he was actually forming part of an illegal reception party.

83

The small sickle of the crescent moon swam low on the horizon among milky-grey clouds, which only allowed the light of a few stars to penetrate, and diffused ineffective rays through the misty dampness of the night air. A small depression on the lee-side of one of the sand-dunes gave Wurr, the Naval Commandant and myself some protection against the penetrating wind, and ensured the best conditions in which to listen for the gentle hum of a distant aircraft. I visited the isolated groups round the edge of the area every quarter of an hour—the ORPO detachment, the SIPO, and Willy, who sat with F2087 in a small car which was carefully hidden from sight close to the area. F2087 was to have his first conversation with Abor in this car.

When I visited the SIPO group at about 0100, Schreieder remarked, with simulated regret and a slightly mocking undertone to his voice, that he had gladly sacrificed the night's rest of himself and his men, but that we should soon be thinking of giving up. His young superior, Wolf, murmured something about the ill-effects of cold, damp night air, and the risk to the health of his staff. London could well bring us out here night after night without anything ever coming of it. . . .

"The time arranged is between midnight and 0200, *meine Herren*," I said. "I would even add an hour for luck, as I have so much confidence in the Tommies' arrangements. Accidents are, of course, quite on the cards, and quite impossible to forecast. We haven't enlarged to the other side on the dangers of engine trouble, crashes, night-fighters, flak, or simply losing their way. But if we are spared such mischance, we'll see Tommy and his load all right."

Stumbling over the Naval Commandant's driver, I told him to take a bottle of rum over to the SIPO party. We had foreseen such a requirement for the cold of early morning.

Soon after I got back to our sheltered corner, I thought I heard a distant humming, but once more it was just the noise of the wind in the grass and the undergrowth. The quietest night and the loneliest moor have their noises, too. . . . I thought of the many nights I had spent during the First World War in fields and moors, in wastes of sand or snow, in rocks and mountains—but always under the good protection of Mother Earth. But never as now had the impending seemed so unknown, the adversary so stealthy, the prize so great. Had they discovered our game? Were they perhaps on their way to our lighted triangle with a load of bombs, to plough us into the moorland with a row of heavy explosions? Everyone had orders to remain under cover from the moment the aircraft appeared until the drop, except for the men manning the lights. These, if bombs were dropped instead of the promised containers, it would be impossible to protect.

I interrupted sharply a murmured conversation between Wurr and the Naval Commandant.

"Do you hear anything?"

We listened attentively. There it was again, louder now, the gentle, far-off hum of aircraft engines.

"Ready, Wurr? That may be it. Everyone to their stations!"

Schreieder bobbed up. "What's up? We can't hear anything."

"Come in here," I said, drawing him behind our cover; "you'll hear better."

We strained our ears. There was no doubt about it now— the noise was coming closer and closer. Still at a great height, but growing steadily in volume. With the large white lamp under my arm I went with Wurr to the apex of the triangle and spoke once more to Heinrichs and his men.

"When I show a white light, you switch on 'Red' in the direction of the aircraft, or the direction of the engine noise, until the drop has taken place. When you see the drop, come here to me at once, but take care that you don't run into an agent on the way. If you didn't answer him in Dutch there might be shooting."

As the white beam from my lamp shone up into the air, the rapidly increasing sound of the engines showed that the aircraft was heading for the triangle of light. Suddenly a shadow drove past us overhead, silhouetted against the pale moonlight. As it did so it switched on red and green navigation lights. I estimated its height at 1,500 feet. After a wide sweep, during which it lost height sharply, the aircraft turned with roaring engines, and now at hardly more than 600 feet, directly towards the apex of the triangle. . . .

The decisive moment had arrived. Containers or bombs—

we'd know any moment now. Suddenly directly above us there appeared in the wake of the aircraft several dark spots which grew large and black as they descended. Gigantic black shadows rushed earthwards—as five large parachutes bore down their loads. Four heavy containers hit the ground at the base of the triangle with a dull impact, while the parachutes followed noiselessly after. This, then, was our answer!

The aircraft, a twin-engined bomber, had gained height immediately after the drop. Its navigation lights blinked twice in salutation as it disappeared westward into the mist.

Wurr, Heinrichs and I pressed each other's hands dumbly—Heinrichs had made his way to us noiselessly from the base line. There were more congratulatory handshakes when we came up with Schreieder and Wolf in their hiding-place. The rivalry and mutual distrust between the Services had vanished before the overwhelming impact of this event.

During the eighteen months which followed we arranged nearly two hundred more drops of men and material by radio with the Anglo-Dutch Secret Service in London (the MID, or Dutch Military Intelligence Service, was attached to the British SOE, or Special Operations Executive), which directed these operations. And we were destined to spend many nights on moor and heath, in radar stations and in the area headquarters of the night-fighters. My memory over this period recalls a broad and anxious stream of secret activity, of wearisome operations and a remarkable degree of self-fulfilment. A few islands still stand out from this stream, and one of these is the night of that first drop at Steenwijk. It left an indelible impression and it was the first roll of drums which introduced the drama that was to follow.

The meeting of Abor with F2087, beginning with the password, had led to a long conversation inside the small car hidden nearby, which was to obtain all the information worth knowing, as a prelude to Abor's subsequent interrogation. At about 0300 F2087 suggested that they should set about burying the parachute harness and remaining equipment so that they could get away in the morning twilight. As he stepped out of the car, Abor was arrested by the SIPO, who took him straight to The Hague. The material, containers and parachutes were loaded on to a lorry and removed to the Abwehr headquarters in Scheveningen. By 0400 the dropping area lay quiet and deserted once more.

Meanwhile, F2087 should have been dictating a report to Willy at the house of the Naval Commandant about his conversation with Abor. But when we got there it became clear that F2087 was decidedly the worse for drink and that he must have interviewed Abor while in this condition. His report was consequently inadequate and full of gaps, and would have to be amplified by interrogation. I decided, as a result of this scandalous behaviour, not to use F2087 again at the dropping point, and I handed over this part of my responsibility to the SIPO, who would arrange for Dutch police officials attached to the SIPO to question incoming agents in the guise of members of the Underground.

Operation Watercress, and the agent dropped during it, whose name turned out to be Baatsen, had little influence on the further development of Operation *Nordpol*. Abor, a pleasant-looking young man, was used by the SIPO as an advertisement to show off their efficiency in seizing agents out of the air. This exhibition reached its climax when Himmler visited Holland some months afterwards and Abor was produced for him. Scandal had it that this was brought about by Abor's strikingly Nordic appearance. What these two gentlemen discussed during their interview has not been revealed.

Abor was of no interest to us from the radio point of view, since he was simply to have been met by Ebenezer and from then onwards had to make his way independently as an agent. Subsequent interrogation did not succeed in fully revealing the nature of his duties. Directly after his arrival we, in any case, sent a signal to London reporting the success of Operation Watercress, and confirmed that Abor was well and in safe hands. . . .

Before I continue the story of *Nordpol* I must touch on another affair which occurred in March–April 1942. In March London informed Ebenezer that a certain "Brandy", a ship's captain in Zoutkamp, Groningen province, was still owed five hundred guilders by the British Secret Service. Brandy was a British agent who had had no pay since May 1940. Suspecting that London might have something particular in mind, I sent F2087 to Brandy with instructions to greet him from Taconis and pay the five hundred guilders. Several weeks afterwards London told Ebenezer that Brandy was to arrange for the reception of weapons and sabotage material on the Engelmann-splaat, a tiny islet with a rescue hut on it in the Wattenmeer of the Lauwerszee on the north coast of Holland. The material would be delivered at night by an English MTB, and Captain Brandy was to take it over on behalf of Taconis. To ensure the success of this operation Brandy was to provide a sketch showing the position of the buoys marking the channel through the minefields.

We accordingly started preparations to give the enemy MTB a warm reception. A certain Kapitän-Leutnant B—— was acting as Naval Intelligence Officer at the Abwehr headquarters, Netherlands, and was an old and experienced seaman. B—— seemed to be the right sort of man, and I took him into my confidence over my plans. Thanks to his good relations with the Chief of Staff at the headquarters of the Admiral, Netherlands, at Scheveningen, B—— succeeded in obtaining a chart of the buoyed channel. We passed these positions to London through Ebenezer, coded in the numeral system used for indicating dropping points, and we confirmed that Brandy would pick up the material. For some reason unknown to us London didn't go ahead with this plan, and all that happened was that the swept channel now had to be changed.

Nevertheless, we let Herr Brandy go in peace, and this paid a dividend, since in the summer of 1943 he informed us in perfect good faith that he was acting for the British Intelligence Service as a "post office" on the courier route to Sweden, and we acquiesced with the British wish to set up this "post office" in Delftzijl in the hope of capturing secret correspondence and espionage material. London soon afterwards instructed us by signal to send on a parcel containing unexposed films. They evidently wanted to check whether the parcel would arrive at its destination unopened, as a try-out for their agent and the couriers. But for some reason no one turned up to collect it. Perhaps the people in Stockholm were more careful and less gullible than our employers in London.

In the first weeks after Watercress, our expectation that Ebenezer would soon be sent new tasks by London was subjected to a difficult test. We had not yet had much experience of this sort of thing and the quiet interval seemed all the more ominous by reason of the fact that we had incontestable proof that the London Secret Service was carrying out operations in Holland without making use of our "good offices".

The first of these occasions was in early April. I received a report from the Gendarmerie that the body of a parachutist had been found, the man having fractured his skull on landing against a stone water-trough. Investigation showed that the dead man belonged to a group of agents who had dropped in the vicinity of Holten. In our efforts to clear up this mysterious affair we turned for help to the local Luftwaffe headquarters which gave out daily reports in map form containing details of all enemy air activity during the previous twenty-four hours. The information on which these maps were based was provided by air-observation posts and radar stations, which plotted the course, height, circling positions, etc., of all single aircraft flying across Holland. We were agreeably surprised by the completeness and accuracy of this information. We found, for example, that details of the operations over Hooghalen and Steenwijk on 28th February and 27th March had been fairly accurately recorded. And we were now able to confirm that the dead agent and his companions must have been dropped near Holten on 28th March. Through the Luftwaffe headquarters in Amsterdam we arranged for closer watch to be kept so as to establish the course of single aircraft, which we described by the word "specialists", as accurately as possible. The evaluation of these daily reports, whose accuracy steadily increased, gave us a useful line on the operations which the Allied Secret Service in England had started without our knowledge. Another indication of secret enemy activity came from Funk Abwehr and the FuB headquarters, to the effect that a new transmitter had been heard in the Utrecht area, whose radio-link had been fixed by D/F as lying close to London. Intercepted traffic indicated that this was the same station as that with which Ebenezer worked. And to add to it all Heinrichs came to me in the second half of April with the news that Radio Orange was once more passing "positive" and "negative" signals.

From all this we concluded that at least one group of agents was working in Holland outside our control and that preparations for further drops had been made. All this made me

very uneasy about our play-back on Ebenezer. Had London
smelt a rat?

On 20th April Ebenezer received instructions to collect
material which would be dropped in the previous area near
Steenwijk. I was pretty sure that it would mean bombs this
time instead of containers, so I took full precautions. I
borrowed against the day of the drop, which was 25th April,
three motorised 3.7-cm. flak guns from Hauptmann Lent, the
celebrated night-flier and Commandant of the airfield at
Leeuwarden, which on the day of the operation were sited
round the dropping area after dark. I had the red lights of
the triangle fixed on posts so as not to endanger personnel,
and arranged things so that they could be switched on from
a point 300 yards distant under cover. The same was done for
the white light. The flak battery had orders to open fire in the
event of bombs being dropped, or if I should fire a red rocket.

We switched on the lights as the British aircraft made its
approach at about 0100. "Tommy" flew several times across
the area, but clearly missed his direction, as the lights were
not being pointed at the aircraft. As he crossed the third time
I went to the apex of the triangle and shone my white light
at him until he turned on his correct course. I have to thank
the absence of bombs for my ability to go on telling this story.

This drop was definite proof that London had not yet discovered our control of Ebenezer. I forgot, in my delight, the
lamentation of the young officer in charge of the flak, who had
not been able to fire, and who might never again have such a
prize held in his sights at a range of 200 yards.

The development of *Nordpol* reached a decisive stage at the
beginning of May. All that we had achieved hitherto could
only have been maintained for a short while had not luck,
sheer chance and ingenuity caused to fall into our hands all
the lines by which the London Secret Service controlled MID-
SOE in Holland at that time. At the end of April London
found itself compelled to join up with one another three
independent groups of agents and one other isolated individual.
Since Ebenezer was included in this link-up, we very soon
succeeded in identifying the whole organisation, and finally in
liquidating it. The principal part in this was played by

Schreieder, who kept his nose to the trail until the last of the men concerned had been arrested.

It happened in this way. In the period February–April 1942 MID-SOE had dropped three groups of agents in Holland, each consisting of two men and a radio set. We knew nothing of these operations. Another single agent had been landed on the Dutch coast by MTB. The operations consisted of the following:—

Operation Lettuce. Two agents, named Jordaan and Ras, dropped near Holten on 28th February 1942. Jordaan was radio operator and was to work in accordance with Plan Trumpet.

Operation Turnip. On 28th February 1942 Agent Andringa and his operator Maartens were dropped near Holten. The set was to be operated in accordance with Plan Turnip. Maartens had an accident and it was his body that was found near the water-trough.

Operation Leek. Agent Kloos with his operator Sebes dropped on 5th April 1942. The set was to have been operated in accordance with Plan Heck, but it was rendered useless by damage during the drop.

Operation Potato. On 19th April 1942, Agent de Haas, using cover-name "Pijl", landed by MTB on the Dutch coast. Pijl had no radio transmitter, but was equipped with a radio-telephony set capable of working at ranges up to five kilometres. He had been sent out from London to contact Group Ebenezer.

Since the Turnip and Heck sets could neither of them establish communication with England, these agents made contact with Group Lettuce, which was operating the Trumpet set, in order to report their mishaps to London. It was not clear whether or not London had told Lettuce to establish these contacts. A signal from Trumpet, intercepted on 24th April and subsequently decyphered, indicated that Trumpet had been in contact with Agent de Haas from Operation Potato, but that the latter had been unable to get in touch with Ebenezer. London thereupon ordered Ebenezer to make contact with Trumpet by a signal passed to the radio set under our control, and the circle was complete.

In our judgment London had felt itself compelled to make this fateful link-up through the loss of transmitters Turnip and Heck, and because of Trumpet's report that de Haas had been unable to contact Ebenezer, which was to pass his signals to England.

A loose contact between different groups of agents had the disadvantage from our point of view that imminent arrests could be quickly reported to London, thus making it difficult to play-back a captured transmitter. But if this contact became a close one, as in the present instance where Trumpet was operating for three other groups, the danger for all of them became very great should one be discovered and liquidated by the German counter-espionage. It was highly unfortunate for London that our controlled station Ebenezer had been ordered to make these contacts just at the moment when the groups which were still working at liberty had been linked up directly with one another. I do not know all the details of how Schreieder and his section in a few days achieved the liquidation of the entire enemy MID-SOE network operating in Holland at that time.

Without doubt lack of experience and gullibility played an important part on the other side. The agents were really amateurs, despite their training in England, and they had had no opportunity to work up through practice to the standard required for their immensely difficult task. Generally speaking they could not have reached the standard of a specialist such as Schreieder.

Afu Trumpet had fallen into our hands complete with signal plan, operating and cypher material. The operator Jordaan collapsed when he discovered the extent of the disaster. He was a well-educated young man of good family, perhaps not developed or tough enough for the most dangerous of the jobs known to Secret Service—agent operating. But that wasn't his fault! Jordaan soon developed confidence in Huntemann and myself, and took the chance which we offered him of operating his transmitter, after we had succeeded in getting him through the nervous crisis which followed his transfer to Scheveningen. On 5th May we used Trumpet to open up a second radio-link with London and passed a signal proposing a new dropping area for this group, which we had found a few kilometres north of Holten. The line of communication developed smoothly, and evidently gave London no grounds for suspicion, for the dropping area was approved shortly afterwards, and we accepted the first drop there about a fortnight later.

A third radio-link with London was established in the following manner. The signal plan for Turnip belonging to the dead operator Maartens had been found on the person of the arrested agent Andringa. We signalled to London via Trumpet that Andringa had discovered a reliable operator who would be able to carry out Turnip's signal plan using Maartens' set, and London gave him a trial transmission so as to test the efficiency of this new recruit. The ORPO operator who took the test must have done it excellently, for the next signal from "over there" told him that he was approved. But we soon had new troubles, which worried me a lot.

About the middle of May Heinrichs reported anxiously to me that he and his men suspected Lauwers of having transmitted several additional letters at the end of his last routine period. It was in fact normal to put a series of so-called dummy letters at the end of signals, and his "overseer" had consequently not immediately switched off the set. His mistrust had, however, been aroused. Heinrichs could not himself be present during every transmission by Lauwers or Jordaan, and he requested urgently that the two operators should somehow be replaced by his own men. I saw the overseer concerned at once. The man declared that he did not know exactly what extra letters Lauwers had transmitted, but that they had had no meaning. The man knew quite well that any other answer could have brought him before a military court for treasonable negligence, but since nothing could be proved one way or the other we had to await London's reactions.

I brought in Huntemann to try and find out what had actually happened, as he was on very good terms with both the ORPO men and Lauwers. It emerged simply that Lauwers had made some of the ORPO men much too trusting, had "softened them up" as we put it. The routine periods had become much too comfortable, and the good treatment I had ordered for the operators, with coffee and cigarettes, had broadened into a friendship which was proving highly dangerous. While awaiting London's reactions, I did not tell Lauwers that our suspicions had been aroused. Nevertheless, although there were no clear indications of treachery, we soon afterwards put an end to the operating of Lauwers and Jordaan by once more using the trick of proposing a "reserve" operator—which was immediately approved.

We were now in a position to bring an ORPO man on to the key in place of either operator without London suspecting anything. The instruction and employment of reserve operators drawn from the Dutch Underground must have been quite understandable to them, as it was always possible that a mishap might occur to the No. 1 operator at any time. Profiting by these events, we did not in general use agent operators any longer. After the arrest of agents sent across later on their sets were operated from the outset by the ORPO without any turn-over period. In this procedure we ran the risk that the "handwriting" might have been recorded in London (on a steel tape or gramophone) and that a comparison might easily give rise to suspicion. By means of touch, speed of operating and other individual characteristics of a transmission technique an experienced ear can detect the difference between different operators when on the key in exactly the same way as a musical ear can detect differences between the renderings of different masters.

If the radio organisation of MID-SOE had observed proper security precautions we should never have been able to introduce our own ORPO operators. But since our experience hitherto had not disclosed any special degree of watchfulness on their part, we took the risk. The carelessness of the enemy is illustrated by the fact that more than fourteen different radio links were established with London for longer or shorter periods during the *Nordpol* operation, and these fourteen were operated by six ORPO men!

I must now touch on certain Dutch internal affairs which were connected with the radio side of the operation, but which had no influence on the main course of events. These were our experiences with "George" and our attempt to carry out the task given to the agent de Haas, cover-name "Pijl."

Towards the middle of May London told Trumpet to contact a man named George. George was quite unknown to us, and interrogation of the arrested agents had failed to establish his identity. As we guessed that he must be an agent who was still at large we had to try and lay our hands on him. If we failed, the operation would be in considerable danger, to say the least. The first difficulty was that the meeting-place indicated by London for George and Trumpet was so cunningly disguised that we could not establish where it was without further particulars. A question back to London could easily arouse suspicion, and we had no alternative but to try and puzzle it out. Finally Heinrichs came to the conclusion that a public-house in Amsterdam was referred to—the Bodega in the Leidser Poort on the Leidschen Plein. We passed our solution to the SIPO and asked them to do what was required.

At the time arranged the SIPO put the agent Andringa at the Bodega in company with several SD men dressed as ordinary habitués, hoping to arrest George when he appeared. But a number of visits produced no results, and all we could then do was to inform London of this negative outcome. A little later London told us to cease our attempts to make contact, probably because we were exposing ourselves in the search for George. We should certainly have been in considerable danger had the good George been by chance arrested by the Germans in the meantime. . . .

I only learnt later that the brave "Akkie", which was the cover-name for Andringa, had done a very plucky thing during this affair. George had actually been present during one of the visits to the Bodega, and Akkie had succeeded in warning him, unnoticed by his SD escort. George was able to escape, and afterwards succeeded in reaching England, but too late to be able to do any harm to Operation *Nordpol*.

And now the story about Pijl. In Operation Potato, Agent de Haas, cover-name Pijl, had been landed secretly on the Dutch coast on the night of 19th April. The circumstances were as unusual as were his task and his equipment. After an adventurous journey by MTB through the German mine-fields and patrol craft he had paddled ashore alone in a rubber boat with all his belongings. Pijl lost no time in carrying off his heavy box with him through the sand-dunes. A fortnight later he fell into the hands of the SIPO through the liquidation of Groups Trumpet and Turnip. When arrested he was in possession of a new kind of radio-telephone set and a signalling lamp of a type unknown to us, and we questioned him closely on the object of this equipment. Pijl was a tall, fair man of about twenty-five, and his frankness was disarming. He stated quite simply that the radio-telephone was for communication with English ships at sea, and that the signal lamp, when switched on, emitted an invisible infra-red beam which indicated its exact position on the shore to a receiving apparatus on board the ship. His task was to seek out suitable points along the coast where more material and agents could be landed at night. He was to arrange with London through Ebenezer the position of these points and the arrangements for supply. He would inform the supply boat by radio-telephone that the coast was clear, and the infrared lamp would indicate the exact landing point. . . .

A well-constructed plan, and very useful equipment which could have caused us a lot of trouble if in the hands of tough and resolute men. Pijl was neither one nor the other, apart from which he had had very bad luck.

Pijl's capture, together with his equipment, brought up new and interesting possibilities and we did not delay in making use of them. We therefore selected suitable points on the coast, cyphered them up by means of the table used for dropping points, and passed them to London over Ebenezer. After a certain amount of discussion, Katwijk was finally approved from "over there". A radio-telephone test was to be carried out during the next moon period; the signal for the operation would be through the usual "positive" and "negative" numbers, and the time was fixed for between midnight and 0200.

It was a cool, starry, moonless night this time as Wurr, Huntemann and I arrived at the bandstand on Katwijk pier. Heinrichs, who had already mastered the intricacies of the radio-telephone, was waiting for us there together with the SIPO, who had Pijl with them. The R/T set, a combined transmitter and receiver with a small fixed aerial, was able to work over distances up to five kilometres, quite a good performance for the portable transmitters of that time. When working, the set was slung across the chest, the dry batteries, which were interchangeable, being held in a broad belt. Pijl began to operate about midnight, that is to say he switched on the set and began to count slowly up to a hundred, then repeating the process—the method agreed with London. Pijl counted in Dutch. We had a second pair of headphones handy for use when communication was established, and the infra-red lamp was in readiness to turn its invisible beam in the direction of the boat when requested.

I had intentionally not kept the German Admiralty informed. The enemy boat must find the situation exactly as it would be on a real occasion. Pijl counted and counted. Every now and again a short rest, a few puffs at a cigarette, and then on again. Time passed, but there was no sign that any boat at sea had heard his numbers, and no reply was received to them. After 0200 we packed up, and in the morning we signalled to London that Pijl had carried out the trial as arranged, but without result. No explanation was forthcoming from the "other side," who merely sent a request that we should prepare for another test in a few days' time. This time it was to be more eventful!

The night was again moonless, but on this occasion very dark, for a completely overcast sky hid the stars. Phosphorescence such as I have never seen turned the sea to magic, the crests of the waves and the surf glittering like fireworks, at times resembling brightly lit ships moving broadside on towards the shore. It was already an hour past midnight, with Pijl counting and counting, but nothing whatever had happened. Suddenly there broke out a furious burst of fire off the shore from Katwijk, as if it had fallen from Heaven, and the bright flashes of quick-firing guns, brilliant patterns of tracer-ammunition and the roll of gunfire told us that a fight was in progress between ships on a westerly course. If that was "our" Tommy, we could go home with a good conscience, as he wouldn't be coming today.

Next day we reported our night's experience to London and enquired what had happened. A few days afterwards we were simply given the disappointing information that the trials were cancelled for the time being. Evidently someone had burnt his fingers on the German sea patrols.

I heard later from our Admiralty that a German minesweeper had been in action with an enemy MTB.

In accordance with MID-SOE's plan, Pijl's duty from now on

was to assist the Ebenezer group and work on sabotage and demolition of German ships in Dutch harbours. During the whole course of our operation nothing struck me more forcibly than the strangeness of this business of carrying out the enemy's operational tasks by means of my own reports covering the activities of dozens of agents—who in actual fact were all the time sitting contentedly behind bars. I will give some examples of this later on.

The first increase in the number of arrested agents had been at the beginning of May, and I had prevailed on Schreieder at that time to have all the *Nordpol* men accommodated in the Wehrmacht prison on the Pompstationsweg at The Hague. Liaison was much better there, and certain dangers were avoided due to the fact that the officers and men at the prison, as well as all the remaining prisoners, were themselves members of the Wehrmacht. I heard from the Abwehr headquarters that my request regarding the penalties to be imposed on the *Nordpol* agents had been forwarded to Reich Security Headquarters (Reichs-Sicherheits-Hauptamt) with a strong supporting recommendation. The Abwehr headquarters and OKW were not in a position to make an independent decision, since under the Civil Government in Holland the RSHA held overriding authority in police matters.

In June I received a message from the head of the Sicherheitspolizei at The Hague informing me that the RSHA had given its formal assurance that the agents who fell into our hands through Operation *Nordpol* would not be given the maximum penalties. I was still at that time confident that written assurances from the RSHA meant that the lives of the agents were completely safe. Lauwers and Jordaan, to whom I had passed this decision through Huntemann, seemed very pleased. The news restored to them and their companions some of the composure and balance which men require in order to overcome a long period of captivity without mental disturbance.

The expansion of the Ebenezer link into three controlled lines, which shortly afterwards became six, resulted in my reports being viewed with unconcealed distrust by "experts" as regards the genuineness of the play-back. It had hitherto been unknown for a line to be controlled continuously over a period of three months. When, from June on, the operation expanded to an unbelievable extent the doubts of these experts were set at rest, at least so long as a stream of supplies and drops continued. Outsiders who had to be brought in from time to time usually took this to be a grotesque figment of the imagination of a poor fellow who had clearly suffered mentally from the burden of his supposed responsibilities. In any case what could the pinpricks of a small nest of secret agents in London effect against the Atlantic Wall which was now being built, the ever victorious German Wehrmacht, and the greater German Reich? This was made fully clear to me in an interview which the Chief of the General Staff, General Schwabedissen, had arranged for me to have with the C.-in-C. Netherlands, General Christiansen. When I had described our experiences and the results we had achieved from *Nordpol* to "Krischan", the old sailor looked at me, nodded his head, winked, and observed that I could tell a very good story. His face showed that "lie" would have been a better word. He hadn't believed a word of it! So it was my duty to report the true state of affairs without exaggeration to high German authorities —who were more than dubious—and at the same time we had to feed the enemy with fairy-tales—which were believed!

We should soon know whether London would continue with the existing arrangement of dropping agents "blind", that is without a reception party, or whether, as in Watercress, they would pass material and agents in through us. From the statements of captured agents we knew that at least eight men had recently finished their training in England and would soon be ready for operations. The blind method of dropping had resulted in losses, for instance Maartens, and we waited anxiously to see which method London would now use.

In fact it turned out to have been decided not to drop any more blind but in future to send them through the contacts already existing, who would arrange to meet them.

In the course of the spring we had amassed a considerable store of knowledge about the enemy's plans, his methods of operating and his radio and cyphering systems. With the help of this experience we could probably even have dealt with blind drops had any more taken place. If the enemy had discovered the truth at this time, he would have had to re-build a difficult, costly organisational structure, employing entirely new methods. Even making allowance for the fact that MID-SOE had not the slightest suspicion of the true state of affairs, it is a fact that the decision to drop "by arrangement" was the chief reason for the catastrophe which followed. This arrangement, which was carried out rigidly and without variation for over a year, was the really dramatic feature of *Nordpol* amid the many other mistakes of omission and commission made by our enemy.

One single control group, dropped blind and unknown to us in Holland, with the sole duty of watching drops which had been arranged, could have punctured in an instant the whole gigantic bubble of Operation *Nordpol*. This unpleasant possibility was always before our eyes during the long months of the play-back, and it kept us from getting too sure of ourselves. We could never forget that each incoming or outgoing radio signal might be the last of the operation.

The decision of MID–SOE was confirmed when the period from 28th May to 29th June brought three dropping operations, for which the "reliable" groups Ebenezer and Trumpet had to provide the reception parties. The operations were:

Operation Beetroot (via Ebenezer). Agents Parlevliet and van Steen dropped near Steenwijk. Duties—to instruct in the Eureka apparatus, guiding beacons for aircraft. Radio communication in accordance with Plan Swede.

Operation Parsnip (via Trumpet). Agents Rietschoten and Buizer dropped near Holten on 22nd June. Duties— organisation of sabotage in Overijssel. Radio in accordance with Plan Spinach.

Operation Marrow (via Ebenezer). Agents Jambros and Bukkens dropped near Steenwijk on 26th June. Duties— organisation of armed resistance in Holland. Radio in accordance with Plan Marrow.

The duties prescribed for parties Beetroot and Marrow were of such importance subsequently that I will discuss them in detail. The Beetroot party was welcomed on its arrival by Underground representatives who were in fact Dutch police working for the SIPO. The arrests were made after dawn, by which time the reception party had had time to find out what the duties of the group were to be. Actually this plan broke down in the case of Beetroot, but was highly successful in all the remaining cases. On subsequent occasions we often discovered important details from the enemy's side, particularly about their secret operational intentions. For example, a single operation gave us precise information about the agent radio schools in England, including the numbers under instruction, their nationality, the teaching staff, standards of ability, etc. Later on our knowledge extended into an accurate picture of the inner circle of leading personalities "over there".

The evaluation of this intelligence, which was arriving continuously and at first hand, gave in the course of time a complete picture of the enemy intentions, since a large number of these agents, Poles and others, who were sent to the Eastern theatre, also passed through these schools in England. SOE organised the secret war in Poland and the risings behind the German lines using agents dropped by exactly the same methods as they intended to use in Holland.

The initial nightly conversations after meeting agents, in which they had to be convinced of the fact that they were talking to prominent leaders of the Dutch resistance, obviously lightened our task considerably. Subsequently these agents had to be given an overwhelming impression of the omniscience of the SIPO, which was conveyed through the introduction during their interrogation of countless details touching the inner circle of the Allied Secret Service in London. We soon got so far that the SIPO could allude to the existence of direct contact with traitorous elements inside the headquarters of their London employers. Although no such elements ever actually existed before, during or after *Nordpol*, it was not difficult to persuade demoralised agents of the truth of this story by using our accurate knowledge, and so convincing them that traitors in London had betrayed them into our hands. In these circumstances it is understandable that the feeling of having been betrayed should have led to frank and truthful statements which gave us important information.

In the meantime Parlevliet and van Steen (the Beetroot group) had remained silent. They were Dutch gendarmes who were not yet willing to fall into our trap. The first conversation before their arrest had failed to reveal the object of the gear which they had with them. This was carefully packed in sorbo rubber in two large boxes dropped by special parachute, and examination by the FuB headquarters had shown it to consist of some kind of transmitter which emitted a continuous signal on a fixed wavelength. It was a so-called "beacon" for aircraft, the general use of which was known, constructed in portable form for the use of agents and fitted with a demolition charge for destruction in emergency.

A few days after the Beetroot drop I was present at an interrogation of Parlevliet and van Steen by Heinrichs, which produced no result, as both men professed to have no knowledge of the apparatus. They started to talk a little, however, when Heinrichs gave them a lecture on the construction and use of the beacon and read out the signals from London which described them as trained instructors in this type of gear.

The beacon aroused considerable interest in Luftwaffe circles, which expressed a great interest in obtaining the counterpart of the beacon fitted in the enemy aircraft. During the spring of 1943 three more Eureka sets were supplied by parachute. This Eureka apparatus had solved the problem of how to direct an aircraft on the darkest night accurately towards a position by enabling it to fly down the beam of the beacon. This kind of pinpointing made the use of lit triangles and light signals unnecessary during drops—a tremendous advantage when operating in enemy territory. Furthermore, drops could now be made independently of the moon period.

We had an impressive test of its efficiency when the gear was used, at London's request, for the first time when dropping Group Heck. This took place at a spot a few kilometres south of Amersfoort. The night was pitch-dark, misty and cloudy, so that you could literally not see your hand before your face. We had set up the beacon's cross aerial in accordance with the instructions and a gentle humming noise showed that it was working. Shortly after midnight an invisible aircraft passed over the area at a height of a few hundred feet—we heard it turn, and then cross its former track and fly very low over us. At the same moment six containers were dropped and struck the ground at a distance of less than three hundred feet from the beacon. The operation had been carried out without the use of any light signals from ground or aircraft.

The reports of these events caused much alarm among the departments of the German Air Ministry concerned. They were only satisfied in the spring of 1943, when the most important portions of the beacon's counterpart were recovered from the wreckage of a four-engined enemy bomber which crashed near Amsterdam.

Group Parsnip, which had been dropped on 22nd June near Holten, had a normal assignment, namely, the organisation of a sabotage group in Overijssel. Parsnip was consequently played back normally by the customary process of opening up communication, agreeing dropping points and accepting drops. It was noteworthy that the operator Buyzer was, on London's orders, also supposed to transmit for Potato (de Haas), Potato having previously worked through Ebenezer. Ebenezer's burden had been lightened in this way because London considered it to be the most reliable of its links and intended soon to use it for an important special task—the blowing up of the aerial system of the Kootwijk radio transmitter.

At the beginning of July London told Ebenezer to make a reconnaissance to see whether the aerial system could be blown up by a demolition commando under Taconis. In a series of signals exact details were given of the method by which the whole system could be destroyed by means of small charges placed at special points among the mast-anchors. I accordingly sent out a reconnaissance party of our people under Willy, who were to conduct themselves exactly as if they were members of the Underground, to find out in what way it would be possible, by day or night, to approach the aerial system, and how the operation could then be carried out. The precise state of affairs as reported by Willy was then signalled to London. We reported a rather small guard, and an inadequate watch over the surrounding area. The demolition of the anchors would not present much difficulty. London signalled back that Taconis must make his preparations in such a way that the demolition could be carried out on the night following the receipt of a prearranged signal.

Towards the end of July we reported that Taconis and his men were ready, and were told by London to stand by, but on no account to start anything before receiving the signal. By the time this signal came I had already thought out reasons for "failure".

Two days later Ebenezer passed the following message to

LONDON CALLING NORTH POLE

London: *Kootwijk attempt a failure. Some of our men ran into a minefield near the anchors. Explosions followed, then an engagement with the guards. Five men missing. Taconis and remainder safe, including two wounded.* And the next day: *Two of the five missing men returned. Three others were killed in action. Enemy has strengthened guard on Kootwijk and other stations. Have broken off all contact. No signs yet that enemy is on our track.* London signalled back somewhat as follows: *Much regret your failure and losses. Method of defence is new and was not foreseeable. Cease all activity for the present. Greatest watchfulness necessary for some time. Report anything unusual.*

A fortnight later London sent Ebenezer a congratulatory message for the Kootwijk party, adding that Taconis would receive a British decoration for his leadership. The medal would be presented to him at the earliest opportunity. . . .

The attack planned on the Kootwijk transmitter was clearly aimed at the destruction of the radio-link by which the German Admiralty communicated with U-boats in the Atlantic. When some days later the English made their landing attempt on the French coast near Dieppe we saw another reason why Kootwijk had been intended to be destroyed. Somewhat late in the day, the German Admiralty hastened to carry into actuality the form of defence for the aerial system which we had conjured up in our imagination.

By arrangement with IC of the Wehrmacht staff, Rittmeister Jansen, I had a reference to the Kootwijk affair published in the Dutch Press. The article referred to criminal elements who had attempted to blow up a wireless station in Holland. The attempt had been a failure, and captured sabotage material had pointed to enemy assistance. The law-abiding population was warned once again against committing or supporting such acts. . . . I hoped that my opponents in London would receive this report by way of neutral countries.

A description of Operation Marrow which follows covers the decisive phase of *Nordpol* from June 1942 until the spring of 1943.

We knew from the first conversations on the night of the drop what the tasks were which had been given in London to the leader of Marrow, Jambroes, and his operator Bukkens, in

broad outline. The plans of MID-SOE, revealed by interrogation, were on a big scale which underestimated the Abwehr
potential on the German side. Typical of this was the lack of
understanding of the true position in Holland concerning the
morale of the population. There is no doubt that the willingness of the mass of the people to participate directly or indirectly
in preparations for underground warfare did not correspond
with London's expectations. It was not until one to two
years later that morale grew gradually more favourable towards such plans as a result of the military defeats of the Third
Reich, the growing Allied superiority and repressive German
actions both against the population and against the economy
of the Western occupied areas.

By the terms of Plan Marrow, Jambroes, who was a Dutch
Reserve officer, was to establish contact with the leaders of
the organisation OD (Ordedienst) and get them to provide
men to carry out the plans of MID-SOE. Sixteen groups,
each of a hundred men, were to be organised all over the
country as armed sabotage and resistance nuclei. Two agents
from London, a group-leader-cum-instructor and a radio
operator, were to take over the leadership, organisation,
training and arming of these groups. No doubt this plan
looked fine from an armchair in London. But its fulfilment
was postponed indefinitely by the fact that Jambroes never
met the leaders of the OD.

It soon became clear to us that we could not play back
Jambroes' task, because as we did not know who were the
leaders of the OD we would not be able to tell London what
Jambroes had discussed with them—when Jambroes himself
was all the time under arrest. So we had to put it to London
that the task originally assigned to Jambroes was impracticable,
and take action in accordance with what we imagined to be the
true state of affairs. We now proceeded to overwhelm London
with a flood of reports about signs of demoralisation among
the leaders of the OD. The leadership, we said, was so penetrated by German informers that direct contact with its members
as ordered by London would certainly attract the attention
of the Germans. When the replies from London began to
show signs of uncertainty and instructed Jambroes to be
careful, we started a new line. This proposed that Jambroes

should make contact with individual and reliable leaders from OD area groups, so as to form the sixteen groups planned by consultation with the middle and lower OD levels. Our proposal met with some objections, but was finally recognised in a practical manner by the increasing of the support through agents and material given to Group Marrow and its supposed component organisations.

The build-up of the Marrow organisation began in August 1942. Naturally at no time were links established with OD groups or with their leaders. On the contrary, we assured London repeatedly that we were making use of more reliable and security-minded individuals. The development of the sixteen Marrow groups had soon made such apparent progress that between the end of September and November London sent across seventeen agents through our hands in Holland, most of whom were destined for Marrow groups. Five were operators with independent radio links. We had these five lines in working order by the end of November, operating in accordance with Plans Chive, Broccoli, Cucumber, Tomato and Celery. Each of these five groups set to work and were soon able to give dropping points to London, which were approved and supplied continuously with materials. At the beginning of December we signalled a progress report of the existing state of the Marrow groups to London. According to this, about fifteen hundred men were under training, attached to eight Marrow groups. In practice, these training detachments would have had urgent need of such articles as clothing, underwear, footwear, bicycle tyres, tobacco and tea. We accordingly asked for a supply of all these articles, and in the middle of December we received a consignment in thirty-two containers totalling some five thousand kilos, dropped in four different areas in the course of one night.

Our information indicated that a new party of agents had completed their training at the secret schools in England about the middle of January, in preparation for action in Holland. From 18th January to 21st April 1943 seventeen more agents were dropped by MID-SOE and met by our reception parties. This time again the majority were group leaders and instructors for Marrow and other sabotage groups. One party of two men had intelligence tasks. Another two-man party was given the task of establishing a courier line from Holland via Brussels and Paris to Spain, and a single woman agent who arrived had been given intelligence duties. The newcomers included seven operators with independent radio-links.

The agents supplied in the spring of 1943 fulfilled the requirements of personnel for the MID-SOE groups which had been planned in Holland. With my few assistants, I was faced with the problem of keeping London's operational maps supplied with information about the multifarious activities of nearly fifty agents, and it seemed impossible that we could keep this up for long. To meet our difficulties an attempt had to be made to get London to agree to a reduction in the number of working radio-links which were now available. We accordingly proposed "for reasons of greater security" to close down some of the Marrow transmitters. These sets, we said, would form a reserve in case some of the active transmitters and their operators should be knocked out by German action. We subsequently arrived at the position where of all the Marrow sets only Marrow I to Marrow V remained in operation.

Although several times between the autumn of 1942 and the summer of 1943 we had reported one of our controlled transmitters as having been knocked out by German action, we had been compelled at times to operate as many as fourteen lines simultaneously. A reduction in radio traffic was essential for the one reason alone that we had a maximum of six ORPO radio operators at our disposal for handling the entire radio traffic with London, and these men were being continually worked up to the very limits of their capacity.

This account of how agents were dropped direct into our arms has not yet described any efforts by MID-SOE to get knowledge of the true state of affairs in Holland. Though there was no lack of trying, these attempts never made allowance for the fact that a possibility did exist that the entire communication network and all the agents sent in were in German hands. The most noteworthy enemy attempt at control, which may perhaps have been one of a number which we did not recognise as such, occurred at the time of Operation Parsley on 21st September 1942. There was little doubt that the agent who was dropped, a certain Jongelie, cover-name "Arie", had a control task. Shortly after his arrest Jongelie declared that in order to confirm his safe arrival he must at once signal to London: *The express left on time.* By saying this he put his SIPO interrogators in a quandary, a situation which they were meeting for the first time.

I had spent the night of the Parsley operation in the dropping area, which lay a few kilometres east of Assen, and had returned to The Hague at about 0700. At nine the telephone bell roused me from my slumbers, and the head interrogator of Schreieder's section IVE informed me of what Jongelie had just said. He added that this message would apparently have to be despatched at the first routine period at 1100.

Half an hour later I was sitting opposite Jongelie in the Binnenhof. He was a man of about forty, with a broad, leathery face, who for a long time had been Chief Operator for the Dutch naval headquarters in Batavia. After a short conversation it was quite clear that Jongelie had developed some Asiatic cunning during his long period of service in Indonesia. With an unnaturally immobile face, he answered my pressing questions repeatedly with the statement that he must pass the message *The express left on time* at 1100 or London would realise that he was in German hands. Finally I pretended to be convinced. Seemingly deep in thought, I said that we would pass his message at 1100—and then, as I suddenly raised my eyes, a gleam of triumph appeared in his. So this was treachery! At 1100 we passed the following message: *Accident has occurred in Operation Parsley. Arie landed heavily and is unconscious. He is safe and in good hands. Doctor diagnoses severe concussion. Further report will be made. All material safe.* Three days later we signalled: *Arie regained consciousness for short period yesterday. Doctor hopes for an improvement.* And the next day the message ran: *Arie died suddenly yesterday without regaining consciousness. We will bury him on the moor. We hope to give him a worthy memorial after victory is won.*

I have related this case in detail as an example of how competent, tough agents, who had been appropriately prepared in London, could easily have forced us into the position where a single treacherous report would have blown the gaff. All we could do in such cases was to pretend that the man was dead or that he had been arrested by the Germans. A series of such "accidents" would probably always have been less dangerous

than the possibility of treachery. Shortly after the Arie incident
London began to press us to send Jambroes, the head of the
Marrow groups, back to London for consultation, Jambroes
having to name a deputy to act for him in his absence. This
request accorded with the man's earlier statements that after
three months of preparatory activity in Holland he would be
required back in England. A reference to the possibilities of
Jambroes' journey was now never absent from our interchange
of signals. At first we described him as indispensable due to
unforeseen difficulties in the building up of the sixteen groups,
and in due course we found new excuses, in which the difficult
and lengthy journey by the insecure courier route into Spain
played the principal part.

1942 went by in this way. At the beginning of 1943 the
requests from London for a personal report became more
urgent and were now broadened to include representatives
from other groups. Innumerable signals passed. London
began to demand information about areas in Holland where
land or seaplanes could be sent to pick up couriers or agents.
We were unable to find suitable areas, or, alternatively, those
which we did find and reported did not suit the gentlemen "over
there"—or else we would suddenly declare them "unsafe",
whenever the organisation of a special flight seemed imminent.

On various occasions we reported a number of agents as
having departed for France, who were expected every month to
arrive, but naturally never did so. Finally we took the only
course still open to us and reported Jambroes as missing . . .
informing London that our investigations showed that he
could not be traced subsequent to a German police raid in
Rotterdam. . . .

On 18th January 1943 Group Golf was dropped into Holland. Golf's duties were to prepare secure courier routes
through Belgium and France to Spain and Switzerland. The
group was well supplied with blanks for Dutch, Belgian and
French identity-cards, with stamps and dies for the forging of
German passes of all kinds, and with francs and pesetas. We
let about six weeks pass before Golf signalled to London that a
reliable and secure route had been established as far as Paris.
The courier for the Golf groups would be an experienced man
with cover-name "Arnaud". In actual fact Arnaud was none

other than my Unteroffizier Arno, who had effected an excellent penetration of the enemy courier routes by posing as a refugee Frenchman who made his living by smuggling jewels. We proposed to London that we should despatch to Spain via the Arnaud route two English flying officers who were living underground in Holland in order to test the reliability of this "escape line". Our proposal was approved, and London confirmed three weeks later that the men had arrived safely in Spain.

Through this exploit, the Golf group and Arnaud acquired much credit in London, and in the spring and summer of 1943 London gave us details of three active stations of the British Secret Service in Paris which were working on escape routes. These were run partly by French and partly by English personnel and had their own radio-links with London. Obviously we did not permit the German counter-espionage in Paris to take action against these stations, once more adhering to the principle that intelligence is more valuable than elimination. My section under Major Wiesekötter now had a clear view of the inner working of these important escape lines, made possible by the well-sponsored arrival of Arnaud in the organisation by reason of a signalled recommendation by London to the stations concerned.

The responsibility for innumerable captures of couriers and espionage material, of incoming and outgoing agents, and of espionage and radio centres in Holland and Belgium during 1943, inexplicable to the enemy Secret Services, must be laid at the door of MID-SOE's confidence in the Golf radio-link, which had been in our hands since the day of its arrival in Holland. In actual fact Golf rendered certain services to the enemy in order to increase this confidence.

We had proved once again the truth of the old saying: "Give and it shall be given unto you." Numbers of Allied flying personnel who had been shot down and had gone underground in Holland and Belgium had reached Spain after an adventurous journey without ever knowing, perhaps until the present day, that they had all the time been under the wing of the German counter-espionage. We continually reported such departures, giving names and ranks, and when the men arrived in safety we noted with satisfaction the growing prestige of Golf. Unfortunately, this kind of action only further whetted MID-SOE's appetite for an authentic report in person.

In one instance, with the approval of MID-SOE, we sent across to London a Dutch sergeant called Knoppers, whom we knew to be working for the OD. Knoppers could undoubtedly tell them a lot about the situation in Holland, and had probably something to report about the resistance movement—but he had not the slightest inkling of *Nordpol*. This expedient, and others, while they did not increase the security of our work, might perhaps lead to new lines when the individual who had been sent across returned to Holland. Sometimes my staff had to play the comedy parts of agents who were "arrested" at the last moment before departure. This had to be done in the case of the supposed successor of Jambroes, whose part was played by my skilful Abwehr man "Bo" in June 1943. His departure and route were arranged in detail by a number of signals between London and Golf and London and Paris. Arnaud convoyed this "chief" to Paris and introduced him into the enemy organisation by the use of a password which had been arranged through London. But as the result of a "random" raid by the German Secret Field Police on a café in the Boulevard des Italiens the "chief" fell into the hands of the Germans just before leaving for Spain. Arnaud and the English operator "Marcel", who were present at the time, "escaped", and Marcel at once gave London an accurate report of the incident, with regretful allusions to the fate of the "splendid Dutchman".

Before Arnaud had had time to report the successful outcome of this comedy back to me in Holland, London in alarm signalled a warning to Golf that the head of Marrow in Paris had fallen into German hands and advised greater caution.

We felt it necessary to give them some comfort "over there". The "chief" had taken nothing in writing with him, and if arrested intended to pose as a deserter from the Labour Corps. . . . The Marrow organisation carried on as usual and in fact received numerous more drops of material. . . .

In another case three of my men, who were listed in London as agents, fell "accidentally" into the hands of a German frontier patrol as they were being driven in a lorry, concealed in empty orange boxes, up the valley from Perpignan towards

the Spanish frontier. The patrol allowed the driver of the lorry ·
to escape, so that he could forward a reliable report on the
occurrence to the controller of the enemy escape line.

It now became necessary to use new methods if we were to
continue to deceive the Secret Service in London about the
real state of the agent groups operating in Holland. These
methods demanded every resource of cunning and inventive-
ness, and included a descriptive picture of the activities of these
agents. The outcome of individual operations, such as that on
the Kootwijk transmitter, was well known to London through
our reports, but, in addition to this, it was essential that the
information obtained by the "other side" from innumerable
sources in neutral countries, etc., should give confirmation of
lively activity by underground commando groups in Holland.
To meet this requirement I had obtained the services of a small
group of German sabotage instructors from the Abwehr
station in Brussels. We carried out a series of dummy demoli-
tions which were timed and located so as to avoid railway
accidents but which were much discussed by the Dutch railway
personnel. In due course we struck a harder blow.

I had promised myself a really good self-advertisement
through the blowing up of a ship in broad daylight in the
middle of the Maas in Rotterdam. With the help of IC at the
Wehrmacht headquarters, whom I took into my confidence,
we obtained a one-thousand-ton Rhine barge which was
loaded with a deck cargo of wrecked aircraft parts in an
isolated section of Rotterdam harbour. The barge was
manned by a German naval crew and prepared for towing
to Germany. One lovely August morning the barge was
proceeding up the Maas in tow of a small naval tug, and when
the two had just passed under the large Maas bridge in
Rotterdam, shortly after noon, there was a sudden dull
explosion. An enormous cloud of smoke rose above the barge,
which had broken its back and was starting to sink. My
sabotage engineers, who were on board in civilian clothes
posing as "air-force engineers", had brought an explosive
charge on board in a haversack without exciting the suspicions of the crew, and the efficient "Bo" had detonated it
coldbloodedly and at just the right moment.

The Harbour-Master's motorboat, "which happened to be

in the vicinity," with me on board in the capacity of reporter for the Propaganda Ministry, hurried to the "scene of the disaster" to save the barge's crew. In the meantime the tug had slipped the tow, and the barge drifted to the shore, its cargo of old aircraft fuselages and wings trailing in the water. The Rotterdamers stood on the banks in their thousands, yelling and clapping their hands with joy—in fact the publicity was a resounding success!

The Harbour Commander, who was a worthy Kapitän zur See, spent a week feverishly interrogating the barge's crew in order to establish the cause of the sabotage—but he never learned the real truth.

Until the autumn of 1942 the saboteurs were given the general task of only undertaking such acts of sabotage as could not at once be identified; there was, however, no lack of mischief in the means provided and the instructions given them. Typical of such penny-dreadful methods were tubes of grease for introduction into the axles of railway wagons, which took fire when the train started to move; magnetic limpet mines with under-water fuses, for attachment to the hulls of ships; stick bombs, to be placed in aircraft by Dutch auxiliary personnel at aerodromes, which were exploded at heights of over six thousand feet by a fuse actuated by the reduction of air pressure; powdered glass for introduction into the underclothing of German soldiers by sympathisers in laundries, which produced inexplicable skin inflammations; steel rods, knives, and pistols fitted with silencers, which could quickly and surely "deal with" German sentries and other unfortunates. One could enumerate many more of these delightful objects.

At the end of 1942 supplies for so-called "secret activity" gave place to equipment designed for more serious work. The tons and tons of the most modern explosives and fuses supplied, the thousands of automatic firearms with enormous quantities of ammunition, and mountains of machine-pistols and machine-guns removed the last of our remaining doubts.

This secret activity had not in fact been controlled by the enemy—for we had "played" it from the outset. We had no option, however, but to take action ourselves as demands for open attack on shipyards, ships and locks became more pressing, action which had to be so arranged that the material damage which resulted stood in its proper proportion to the effect which we believed it would have on the confidence of MID-SOE. I hoped, for example, that we should be successful in the case of the sunken Rhine barge.

Three of our controlled agent groups, which included Ebenezer and Trumpet, were ordered in the spring of 1943 to undertake a peculiar form of activity, and received the following signal at about the same time: *In our opinion the time has come to begin active hostilities against our political opponents in Holland. Report what measures you consider possible for carrying out attempts against leading members of the NSB and other collaborators. When you are ready we will send you a first list of these individuals.* To this I composed three different replies from the groups concerned, which ran approximately as follows: *(A) We have read your proposal in signal X with astonishment. It is not considered to be in line with our proper tasks. We are prepared for any other kind of work but decline to carry out planned assassination. Request reconsideration. (B) We have considered fulfilment of the task set out in signal X. Please realise that to do this may lead to severe German counter-measures and civil war. Destruction through reprisals of our own centres may result. Request your views. (C) Promising opportunities exist for carrying out task in your signal. Time of execution will depend on the individuals concerned. Send us suitable names in order that the necessary preparations may be put in hand.* A few days later groups A and B received the following identical signal: *We cannot accept your objections and misgivings concerning execution of our plans. The methods envisaged have been particularly successful in other countries. We await your reconsideration. We must be able to count on your co-operation.* Group C received a week afterward from London a list of fourteen individuals who were to be killed, including the following: Rost van Tonningen; van Geelkerken, President of the Netherlands Bank, and his deputy van Mussert; Ernest Voorhoeve, Propaganda Chief of the NSB; Engineer Huygens, organisational head of the NSB; Woudenberg, head of the Dutch Labour Front; De Jager, district chief of the NSB at Drenthe; Feldmeijer, head of the German SS; Zondervan, Chief of the WA.

Naturally nothing came of these attempts. We used our well-tried tactics of hesitation and delay, which we based on numerous grounds, continually producing fresh ones. In the end we asked for rifles with telescopic sights under the pretext that the protection of the intended victims was too well organised to be overcome by normal weapons. As these telescopic sights were not sent in subsequent drops, the project came to nothing. At least we were now better acquainted with the methods used by MID-SOE.

Between Christmas 1942 and the New Year Schreieder invited me to a meeting in the SIPO mess at The Hague. I found among the company a tall, pale man with a shifty pair of eyes set in the face of a scoundrel. Schreieder introduced him as van de Waals, and so I had the doubtful honour of making the acquaintance of this gentleman, whom we knew in the IIIF office to be employed as a SIPO informer and "V"-man by Schreieder. There must be something behind this, as I knew well that Schreieder took the greatest pains not to allow his "V"-men to have contact with Abwehr officials.

Schreieder informed me that van de Waals had posed as an English agent, with a radio-link to England, to the National Committee, an illegal organisation headed by Koos Vorrink. He had been instructed to use his set for the exchange of intelligence between the National Committee and the exiled Dutch Government in London. When Schreieder now asked me to arrange to get van de Waals accredited as a genuine agent by the transmission over Radio Orange of a message specially composed by the National Committee, I could not refuse him. This was one of our first requests for credentials which were passed back for retransmission over Radio Orange in order to introduce German agents. To keep London satisfied we based our regular requests for such credentials on the ostensible necessity for these agents to make contact with important individuals in Holland. After a time the SIPO requirements increased, and the credentials passed over Radio Orange were turned into a weapon used by the SIPO against the Underground. At this point I refused further requests on the ground that they were beginning to endanger *Nordpol*.

The SIPO then had recourse to other methods. On orders from Berlin I passed every week the interchange of messages with London in plain language to Schreieder, for his information. There appeared from time to time in these messages requests for radio communications through Radio Orange, which were required in respect of *Nordpol*. SIPO informers now began to use the text of such messages before they were transmitted back to gain the confidence of individuals in the Underground. I was unable to stop this kind of thing, although I continually emphasised the danger it represented to the radio-links. If a call through Radio Orange should be mentioned prematurely by a man who was later discovered to be a SIPO informer the Underground would soon make the obvious deductions. In the circumstances, therefore, we gave as little notice as possible when making such requests to London.

In the case of the National Committee, the call from Radio Orange came through after only a few days and enabled van de Waals to gain the confidence of Koos Vorrink. We asked London later whether we might pass a report from Koos Vorrink by radio, and this was approved. But when Schreieder handed me this report, which consisted of three pages of type about plans for forming a Government after the war, I refused to pass it. It was quite unsuitable for transmission by radio in its form, contents and degree of priority. We agreed finally to pass the report in a much abbreviated form and in twelve parts. After clearing the fifth or sixth part, London told us curtly to stop this nonsense: the matter was of no importance and was simply endangering the operator. We passed the remainder during January 1943 and received a cold reply from London to Vorrink which contained no serious answer to his proposals.

Some time later Schreieder "blew" the National Committee, and had the leaders, Koos Vorrink, Louis Vorrink and Ringers arrested. From what I heard these three were not badly treated and later on they became leaders of the Dutch Labour Party which had been giving some thought to the construction of a post-war government.

In November 1942 I moved the IIIF office to Driebergen, a few kilometres to the eastward of Zeist, and also had the Heinrichs FuB office transferred there. The SIPO IVE section also had a small office in Driebergen for liaison in connection with *Nordpol* affairs.

After it had become clear in the winter of 1941–42 that

the greater part of the German forces would be pinned down for a long time in the east, the German leadership in the west was put on the defensive. A few divisions only, of doubtful fighting efficiency, and some reserve formations, were available. The build-up of the Atlantic Wall was pressed forward, and the Dutch coast included in the system. In May 1942 the Wehrmacht headquarters staff was moved back from The Hague to Hilversum, and the Abwehr headquarters was transferred with it.

I took advantage of this opportunity to detach the IIIF section from the Abwehr headquarters, as the number of Dutch and German auxiliary personnel who worked at close quarters with us in Scheveningen was causing me some anxiety, from the point of view of cover and of the security of our activities. Driebergen had the advantage of a central position, and security was assisted by the wide separation of the houses, well away from industrial areas. The IIIF office operated from there until the end of the war without its function ever becoming known. After the war I discovered from former members of the Dutch Underground that it had been noted down by them as an administrative section of the ORPO. The mistake was understandable, as ORPO officials in uniform were continually to be seen about the place, and my men generally wore civilian clothes when away from the office.

Part of the *Nordpol* radio network which had previously been keyed from positions at The Hague, Gouda, Rotterdam, Nordwijk and Amsterdam was shifted at the same time. The new transmitting points were Eindhoven, S'Hertogenbosch, Utrecht, Hilversum, Arnhem and Driebergen. We took care not to operate a set from one place for too long. We were unaware of the exact state of development of the enemy's D/F technique and always assumed that bearings would be taken from "over there" as a check.

Driebergen had the further advantage that the Commander of the German night fighters in the North-Western Area had his headquarters in the town. I had been in close liaison with his staff since the start of the mass drops in the autumn of 1942. The IC on his staff gave us first-class support in our operations. He was an outstanding airman and all-round sportsman who was always ready to fly me about over Holland in his *Storch* for hours at a time when I was in search of new dropping areas. These could be found much better from the air than from the ground, and areas had to be selected far from dwelling-houses, main roads and military installations. The presence of flak batteries and air observation posts in the neighbourhood was naturally to be avoided.

The majority of the dropping areas which were used during *Nordpol*, numbering about thirty, lay in the moorland area of the Veluwe, a few to the eastward and northward of Steenwijk, and one east of Assen. One of the most interesting, however, was established by London orders on the shore of the Zuider Zee south of Schellinghout, a few miles east of Hoorn. Two agents were dropped by parachute in bathing dresses into the sea at this point, and they had to swim towards light signals shown by us from the dyke. The operation was not a success, as the men got entangled with their parachutes in the darkness and we eventually got them ashore in an exhausted state. The report we sent in resulted in no second attempt of this kind. One of the dropping-points, about ten kilometres from Driebergen, near Darthuysen, was a special favourite of ours, for we were spared the endless journeyings by night, which often amounted to several hundreds of kilometres. Sometimes my small staff was out night after night, as drops were often made in three or four places simultaneously, on occasions even up to six, and during moon periods operations were often ordered from "over there" for three or four nights in succession.

Our reception preparations were arranged in accordance with a fixed programme. It developed into an all-day and all-night affair which held neither romance nor excitement. When a "positive" number was received in regard to a particular group of agents all that we had to do was to pass a code message to the SIPO, the ORPO and the Abwehr station at Hilversum in order to set the machine in motion. The Reception Committee and all others concerned at once knew their meeting point and which dropping area would be used. For example, a code message which read *Visit No. 17 this evening with two ladies* signified that two agents would be dropped that night at dropping area No. 17. At times two of my radio operators were attached to the reception party whose job was to report developments to headquarters at

Driebergen. They would set up their sets, test communications, and then maintain radio silence until the aircraft had been and the dropping operation had taken place. If no aircraft had appeared by the end of the period allowed they would ask whether "specialists" had been reported as on their way, or whether "dismantling" could proceed—in other words, whether the party could now leave the area. The Luftwaffe radar stations gave us this information, and appropriate instructions could be quickly passed to the dropping area.

Our radio network had obviously to be carefully disguised, and we used prearranged signals to cover all imaginable questions and reports. For example:

AAA— arrived, dropped, flown away.
AXX— aircraft arrived, circled the area, but did not locate the dropping point.
AXA— arrived, area not identified, flown away.

As we never knew for certain at any time whether our game had been discovered or betrayed, we never knew what the next operational night would bring. There was no cause to underestimate our enemy after the outstandingly good work of the Allied Secret Services, and especially SOE, in France, the Balkans and Poland. We had always to reckon with the possibility of bombs on the dropping point, and a prearranged signal BBB was to signify "bombs in the triangle of lights". This signal would serve as a warning to all other areas should the enemy plan to massacre us at six different points on one and the same night. . . .

The central radio station passed reports from the dropping points to the night-fighter operations-room near Harderwijk by direct telephone line. I was normally there myself during the more important dropping operations, so as to watch each stage of the proceedings and to release the night-fighters when the right moment arrived. The night-fighter operations-room was a large dimly lit affair, in the centre of which stood an enormous table with a top of frosted glass, lit from below by dim light, on which the Dutch frontiers, the coast-line and the Zuider Zee were outlined on a squared background. When the radar stations got on to an approaching enemy aircraft a glowing red spot would appear, far outside the line of the islands, on the huge, dimly lit expanse of the table-top. Simultaneously a second point of light would begin to dance on the darkened wall of the room against a numbered scale—which indicated the height. "Specialists" could be identified with reasonable certainty, as they used to fly in along one particular course which became well known to us by experience. They would cross the island of Texel, fly on to about the centre of the Zuider Zee and proceed from there to the dropping points. Then quite suddenly the red spot would start to move in circles on the milky glass, once, twice and sometimes even three times while the dancing spot would disappear from the height scale—the aircraft had descended below nine hundred feet.

When we planned to shoot it down a white spot would join the red one at this junction. This indicated a night-fighter lying in wait above the dropping area, its height being from nine to twelve thousand feet. The night-fighter controller used to speak so quietly, clearly and confidently to the aircraft by radio telephone that the two men might almost have been seated opposite one another. When the red spot began to move once more on a straight course a radio signal would come in from the dropping area. The controller would glance at the message form. At the three letters AAA which it contained the fighter would be released to the attack, and he would dive through the dark air, unable to see further than a few hundred feet, but guided unerringly on to his victim by alterations of course passed to him by the man with the calm voice seated before the illuminated table. Slowly and inexorably the white spot would approach the glowing red one as it continued on its straight course to westward. As the two aircraft thundered through the night at a speed of three to four hundred kilometres per hour—two, three, four minutes of unbearable tension—our eyes would be held as though mesmerised by the gently vibrating, forward gliding points of light, which had now approached so close to one another that it seemed as though they wished no more than to coalesce. A glance at the height scale. The two lights are dancing on the wall opposite one another at the same height. The controller's voice sounds very loud in the deathly stillness of the room.

"You should now be able to see him!"

In three seconds confirmation comes through from the night-fighter, and then the order of the controller falls like a stone into a pool:

"*FIRE!*"

Somewhere in the dark night there's a spurt of cannon fire—a burning meteor is outlined for a moment against the sky, and plunges abruptly in a steep curve towards the earth. . . .

The red spot on the table disappears and the illuminated height scale goes dark again: the chrome and nickel of the instruments gleam coldly and dispassionately, and a cold breath of air as though from a tomb seems to rise from this murderous instrument of glass and steel on which one single white spot has remained. The night-fighter has made a wide circle and is now flying straight as an arrow eastward, along the beam of the beacon at his home airfield, Deelen.

Twelve four-engined bombers were intercepted and shot down according to plan, after dropping their load of agents and material, during Operation *Nordpol*, and the German night-fighters suffered no losses during these encounters. We held the advantage of surprise and of accurate guidance by radar and radio-telephony. These interceptions were partly due to our quest for the aircraft counterpart of the Eureka beacon apparatus which had been supplied to the dropping areas, and we continued our search of aircraft which made use of this gear to fly in until one day a repairable instrument was salved from the wreckage of a bomber. The air engineers on the staff of the Luftfahrtministerium were able to satisfy their scientific curiosity and we were thus finally relieved of their incessant demands for a specimen of this mysterious box!

The proportion of twelve bombers lost to a total of about two hundred organised supply flights was not very high. It was, however, increased by losses unknown to us due to other factors, with a consequence that RAF "specialist" flights over Holland were stopped for a time in the middle of 1943. It had been found that the losses in Holland were very high in comparison with similar operational flights in France and other countries. There was, in any case, an obvious difference in the fact that the German night-fighter net over Holland, that is to say over the approaches to the Ruhr, had been completed by the end of 1942, while those over France and other countries had not been fully established.

In October 1943 we received our last drop of materials. By that time the *Nordpol* operation, this duel in the dark, had run for exactly twenty months since the first drop, and our adversaries had fought throughout with bandaged eyes. Just about two years had passed since we had discovered the identity and tasks of the first group, Ebenezer. Over fifty well-equipped and fully trained agents of MID-SOE had been caught in the meshes of the net which the radio-links between the island and the continent had spread for them. The sands of the *Nordpol* operation were now soon to run out. The "agents to report" whom we had been asked to send across to England for so long, whose journeys we had continually postponed by means of a thousand excuses and devices, were already on their way! Two brave men, who had taken part in the operation on our side and who could report from their own experience on what had occurred during these two years, had deceived us! These two men, who knew all our secrets as well as we knew them ourselves, right up to the last agent dropped, had made their way to freedom. I was convinced that they would make their way back to the island in the teeth of a thousand dangers and adventures in order to destroy the instrument of war to which they and their comrades had fallen victims. A race against time had started, a fight for a final success at the last minute.

CHAPTER III: FINALE

It was just before midnight. Willy and I had been lying for an hour covered by blankets and ground sheets on the heap of sand which surrounded the yawning black hole at our feet. A cold, damp night wind had sprung up at nightfall, after a fine October day, and was blowing in great gusts through the pines and the tops of the birches which stood out against the moonlit horizon around the stretch of moor. In the clearing, which was a few hundred paces wide, were pale coloured heaps of sand round four other holes which I had had dug in the form of a triangle, the fourth hole being a few paces in front of the apex. This was a precaution in case anything unpleasant should occur this evening, which seemed highly possible from the general nature of things.

Some unexpected visitors had appeared at the rendezvous in Hoevelaken, where the reception party had mustered, who wished to witness one of these operations. We had driven together to the dropping area, had looked round the arrangements and had gone off to a piece of woodland well away from the dropping point after a few explanatory remarks from me. They were to stay there to await developments. The new head of the Abwehr in the Netherlands, Oberst Reuter, and the recently arrived Chief of the Sicherheitspolizei, Brigadeführer Naumann, successor to Dr. Harster, had arranged to come together to see this operation, the first after an interval of over four weeks, as though they realised that they must make haste if they were ever to have such an opportunity again.

On 31st August, Queen's Day in Holland, two *Nordpol* agents, Ubbink and Dourlein, broke out of the prison in Haaren and disappeared. I had a short report to this effect on the morning of 1st September from Schreieder's office. Soon afterwards Schreieder himself rang up in considerable agitation to give me a seemingly endless description of the measures which he had taken for their recapture. It was clear to me that, through this incident, the bottom had been knocked out of the whole *Nordpol* operation. Even if the fugitives did not succeed in reaching Spain, Switzerland or even England itself, they were at large—though perhaps only temporarily—and would certainly somehow record their experiences since their departure from England and get this report by some means or other back across the Channel. There was little in Schreieder's report to indicate how the break-out had been effected, but my astonishment at its success was only equalled by my admiration for the audacity of these two determined men. I went to Haaren the same afternoon to get further details.

The isolated four-storied building of the former Theological College in Haaren had been used by Schreieder since the summer of 1941 for accommodating important prisoners of the Sicherheitspolizei. A number of English agents were confined in about twenty large cells on the second floor, two or three to a cell, whose contact with the outer world was completely cut off, but who could talk to one another inside the building as well as during the daily exercise period. It was to be expected, therefore, that each one of them would be fully aware of all the details of every individual case, and that it would be possible for any one of them to betray the entire range of our counter-espionage operations if he succeeded in getting away. The large house was well guarded by sentries, and the grounds so well secured behind barbed wire that it seemed impossible for anyone to break out. From the reports which Wurr and I received as we looked over the ground it was clear that the escape had been carefully prepared. The men had lowered themselves after dark a distance of forty-five feet out of an unbarred corridor window into the yard, using strips taken from the sacking of their mattresses as ropes. They had gained the garden, crawled through the barbed wire and disappeared. What could not be explained was how they had been able to get out of their cells through a small fanlight above the door; furthermore, none of the sentries had heard the rattle of the fanlight as they crawled through. There had obviously been grave neglect in the watchfulness of the sentries and in the supervision of the whole organisation, and there was a distinct possibility of silent connivance or even active co-operation on the part of some of the sentries. These were made up of Dutch SS men who were unfit for war service and whose reliability could not be guaranteed, in my opinion, in any circumstances.

The heads of the prison administration, as well as their Chief, Schreieder, were oscillating between hope and fear. If the fugitives could be quickly recaptured the whole affair could be hushed up and a full-dress investigation avoided. Otherwise the senior officers concerned could be assured of painful measures against those responsible. As always in such cases, the anger of the outwitted was visited on the remaining prisoners, and the small privileges which had been given them were withdrawn forthwith. We represented continually to the SIPO that the guards should be reinforced rather than that the remaining prisoners should suffer, and after some time we gained our point, but this led to a sharp conflict between the Abwehr and the SIPO over the treatment to be accorded to enemy secret agents. Co-operation between us and the SIPO officials had been essential to the success of *Nordpol*. This had contributed to the fact that the perpetual and growing conflict between the Reichssicherheitshauptamt in Berlin and the Abwehr headquarters in Holland was not so severe as was the case with the Abwehr headquarters in other countries. It was now all-important that I should give these police officials no pretext to alter the terms of the agreement previously reached between us as regards the treatment of *Nordpol* agents. We had become aware of a new attitude by the RSHA since the arrival of the newcomers such as Brigadeführer Naumann and the SIPO Chief of the FuB section, Kienhardt.

A decided change for the worse had taken place in the general war position during the summer of 1943. In the East our line against the continually advancing Russians had become thinner and thinner after the bloodletting at Stalingrad. Driven out of North Africa, we had maintained a costly front with great difficulty in Italy, and after the fall of Mussolini complete collapse had only been prevented by the fact that the Allies were clearly not seeking a military decision either there or in the Balkans—as they could at that time have done. The trump card of the U-boat war was no longer good, since the enemy radar had learnt to fix the position of our boats, which were being destroyed in large numbers. Night after night the airborne radar of superior Allied air forces was able to fix their targets in Germany with deadly certainty, and in spite of heavy losses the enemy bomber squadrons were wiping out town after town. How could the change come about which might bring victory? When would the miracle weapons arrive, of which the party fanatics talked so much? When would the "genius of the Führer" in which they believed strike the decisive counter-blow? When would an end come, the end so long prayed for, to the sufferings of the masses now living dumbly beneath the reign of mental terror?

And we old soldiers? Did we still hope for "total victory"? Had we ever really wanted this unconditional victory? Had we not fought from the first day because our Führer had proclaimed "the Fatherland in danger"? Were we not now struggling obstinately simply to prevent total defeat? The total defeat of Germany—our Germany, not the Germany of Hitler and the Party!

The night wind had died away after driving the cloud-banks

away eastward. Willy and I went round the area several times under the cold, starry sky. It was already past one and the night was deathly still. Nothing had stirred in the air. We knew from experience that quiet, moonlit, cloudless nights were not favoured by the enemy. If the aircraft should still come, despite the bright unfavourable weather, the possibility of an unpleasant surprise became more likely. I had not mentioned my anxiety to the ORPO officials who were to work the triangle of lights, but they knew all about it none the less. . . .

Half an hour later all was over. The aircraft had made a sudden appearance and had dropped its load during the first run-in, after a long, direct approach made easy by the very good visibility. Six heavy containers lay within the triangle where Willy and I had been operating the light signals from our deep hole.

Once again we had been successful, and I gave no more thought to the possible results of the disappearance of the two agents six weeks previously. In actual fact we had meanwhile signalled to England that Ubbink and Dourlein had gone over to the SIPO and that they would try to make their way across to England in the SIPO's employ. I did not take too seriously this final attempt in the series of deceptive measures taken against the enemy, and I was convinced that this trick would soon be seen through in London. Nor could our latest success give me back my old confidence, and there seemed to be no reason why I should share the loudly voiced optimism of Naumann and his SIPO men, who certainly underestimated our London opponents.

Willy was as silent as I was during the slow journey back to Driebergen. His thoughts, as well, were turned back to that decisive moment when the aircraft, coming in low, had released the six black shapes from its fuselage.

A few days afterwards the Golf group was instructed by London to prepare a safe house in Brussels for two agents who would be arriving very shortly from England. We were not immediately informed how they would be coming or what their duties were to be. We had a *pied à terre* in Brussels at that time through F2087, who had lived there, still controlled by Willy, since 1942. His good contacts with a number of the leaders of the so-called *Armée Blanche*, the largest Belgian resistance group, had produced some good information about their activities.

Willy was now sent to Brussels to prepare a safe and well-disguised house as quickly as possible. I was intending to run this affair so that the new arrivals would remain in freedom in the care of F2087 and would start their work via his contacts. Subsequent action would depend on developments. Having informed London about the arrangements in Brussels, we very soon had some detailed instructions. The two agents, cover-names "Brutus" and "Apollo", were to be dropped into Belgium, and we were given the code word by which they would identify themselves to us in Brussels. Towards the end of October Golf was told to expect Brutus and Apollo within three days, and to assist in passing them from Brussels into Holland.

I had worked hitherto on the assumption that the two men would have contacts to establish in Belgium, and the latest instructions, namely to pass them through to Holland, were unexpected. We failed to see the point of these new tactics of dropping agents into Belgium who were intended to operate in Holland, with the dangerous journey in between, and we awaited developments anxiously. I had detached Willy to Brussels temporarily and organised a link to give rapid communication between Brussels and Driebergen, for control purposes.

Whatever London might think in general about the radio-links with Holland, their confidence in Golf was evidently as yet unshaken. Had they dropped a few control agents "blind" in Holland they would have been much better informed. The R.A.F., which had preferred to transfer these operations to Belgium on account of the heavy losses over Holland, had a rude shock in connection with Brutus and Apollo. Simply because we had no idea when and where these two agents would be dropped, tragi-comedy intervened and caused their transporting aircraft to be shot down by flak in the area Antwerp-Mecheln.

One early morning a severely shaken and bedraggled individual appeared at the house in Brussels, who was clearly most relieved to be once more safe among friends. He showed great confidence in F2087 and gave him a detailed account of his experiences during the night of the crash. He had jumped, with Apollo and the crew, from the burning aircraft, leaving behind his entire equipment, radio set, etc. He knew nothing of what had become of Apollo, but expected that he would get to Brussels as quickly as he could, if he had landed safely.

I used this fortunate accident through which the two agents had become separated to have Brutus sent immediately on to Holland. In order to regain touch with one another they would be forced to use my courier system, and in this way we could keep ourselves well informed about their plans and activities. A Dutch police official, who worked for us from time to time without being given any background knowledge, was commissioned by Willy to pass a man across the frontier from Belgium into Holland. F2087 drove Brutus to the frontier near Baarle-Hertog and handed him over to the Dutch policeman, who took him safely into Holland on the back of his motorcycle. Brutus was accommodated in Maarn with a group of the Dutch Underground with whom my people had links which were completely unsuspected. In this way we had Brutus under direct control, and indeed he had been most impressed by his reception in Brussels, his secret journey on and his accommodation in Holland.

Apollo arrived in Brussels twenty-four hours after Brutus, and as tattered and woebegone as his friend. His account confirmed that of Brutus. F2087 suggested to him at my instigation that he was powerless without a radio set, clothing, equipment and money, and that we therefore proposed to send him straight back to England via Paris and Spain, so that he could return when he had again been properly equipped.

Apollo might have been the first genuine *Nordpol* agent to return to England to report his experiences, and we certainly banked a lot on him. If he had then reported a safe house in Brussels as being reliable there would still have remained a chance of continuing to play-back the Golf transmitter, even though the reports of Ubbink and Dourlein might have put all the remaining *Nordpol* radio-links out of action. We might then have expected Apollo back in Holland after quite a short interval.

We sent a number of signals to this effect via Golf, but London set its face absolutely against our proposal. Apollo was on the Continent, and he must therefore go on to Holland and carry out his task there. Further instructions and his reports back would be passed via Golf. I was gratified to receive this touching proof of London's confidence in the Golf radio-link, although it did not suit me at all to have my plans upset. Apollo kept nothing back from F2087, so we learnt, incidentally, all about his tasks in Holland. They included a reconnaissance of the German defences along the Yssel line, as well as preparations for supplying the civil population of "Fortress Holland" with food. A large part of the cattle population from Overijssel, Drenthe, Groningen and Friesland was to be moved into "Fortress Holland" before the start of the Allied invasion of the Continent. These humanitarian plans led us to the conclusion that "Fortress Holland" would probably be left outside the area of the main Allied attack.

After a fortnight's interchange of messages London gave way to our obstinate demands that they should agree to Apollo's return, and issued instructions that he should be sent via Golf's courier route into Spain. Brutus sat in Maarn meanwhile, writing fiery articles for the illegal press, which he supplied with our assistance. He also composed a long personal report, the highly favourable background of which would be a magnificent advertisement for Golf. Apollo started off for England with this report in his pocket, taking with him our best wishes and hopes of seeing him again soon.

We allowed Brutus to carry on working undisturbed from Maarn, as his relations with the Dutch illegal press we did not regard as either dangerous or important. The reports which he passed to London were quite sufficient for our control purposes. When we learnt subsequently that our deception had been discovered by London Brutus was arrested.

During the first ten days of December London's signals became so dull and colourless compared with their usual quality that it did not need all our knowledge to enable us to guess that the enemy was trying to deceive us in his turn. Hardly any doubt remained that Ubbink and Dourlein had reached their objective. Nevertheless, we made no move, and gave not the slightest indication that we too realised that the great bubble of the agent-network and radio-links in Holland had been finally pricked. We carried on normally for another month, wondering more and more what tricks London might now try to play on us. But nothing occurred which was worthy of mention. The increasing triviality of London's signals could not now deprive us of the heritage of Operation *Nordpol*.

Another escape of three agents from Haaren took place just before Christmas 1943, and this led to considerable recrimination between Naumann, the head of the SIPO, and myself. In his reports to Berlin Naumann had tried to fasten the responsibility for this latest incident on me, on the grounds that I had insisted on the arrested agents being treated humanely. He was not going to lose the opportunity of settling accounts at last with me and the Abwehr IIIF section in Holland. I heard privately from friends in Berlin that the witch-hunt had started and that I had better make use of my good relations with the staff of C.-in-C. Netherlands if I were not to be accused of having endangered Operation *Nordpol* through my negligence. When Rauter[1] wrote to the C.-in-C. with an implied threat unless he should take action against the Abwehr officer responsible, I had no more doubts on this score. Fortunately my senior officer in Holland, the Chief of the General Staff, Generalleutnant von Wühlisch, was not the man to lie down under such an unjust and malicious threat to one of his officers, and I let him have plenty of material for a counter-attack against Rauter and the SIPO. Both von Wühlisch's reports to Berlin and his letters to Rauter sharply rejected the accusations levelled against me, and he left nothing to the imagination in stating bluntly where those responsible should be looked for.

This attack had not been beaten off before new points of difference arose. A conference took place in Berlin in December 1943 between representatives of the Abwehr and the RSHA— a futile affair. The RSHA representative spoke for an hour on the subject of a counter-espionage operation which had been in progress in Holland for the past eighteen months under the code-name of *England-Spiel*, using a series of radio-links with the English Intelligence Service. The speaker clearly believed that his detailed exposition was in accordance with the facts and that he was telling these gentlemen of the Abwehr something new. The Abwehr IIIF representative enquired whether he was perhaps referring to the *Nordpol-Spiel*, and it then

[1] Chief of the SIPO and the SD in Holland.

emerged that the SIPO in Holland had, since the summer of 1942, reported *Nordpol* to the RSHA in Berlin as having been a SIPO operation—it was unfortunate that they had omitted to mention that it was in fact being carried out exclusively by Abwehr IIIF. The state of mind of the SIPO after having been caught out in such a cheap swindle was hardly conducive to further useful collaboration between us.

This kind of friction was simply the accompanying symptom of the struggle in progress at that time for the control of the Secret Services in Germany. Since the summer of 1943, when General Oster, Canaris's Chief of Staff, had resigned—or, more correctly, had been arrested—it had become clear to the initiated that the independence of the Abwehr at the head of the German Military Secret Service was coming to an end. The previous duplication of the Secret Services, which had been so carefully arranged, no longer seemed safe to the potentates of the Third Reich, since it had been established that the leading figures in the Abwehr were opposed to the Hitlerite war and to the measures being taken to change its higher direction. The whispered rumours about suspected Abwehr contacts with foreign and even with Allied Secret Services had not yet hardened into clearcut suspicions of individuals, nor could anything be proved juridically. In any case the maintenance of contact, through suitable middlemen, with other services is one of the fundamental duties of the Abwehr, if it is to do its work properly.

Obviously we in the out-stations knew little of what was going on in Berlin, and that simply from personal gossip when in the capital. It was in this way that I heard in November that it had been decided that our service was to be incorporated into Amt VI of the Reichssicherheitshauptamt from the beginning of the New Year. This sobering reflection had its confirmation in the flow of voluntary retirement then setting in of a number of senior officers who had served in the Abwehr for many years, and who preferred to be transferred to other duties rather than be directly subordinated to the RSHA.

We of the counter-espionage section meanwhile thought out another plan which might serve to forestall this hateful subjection to the Himmler régime. Making common cause with the western section of the General Staff which dealt with foreign armies, and with the IC of C.-in-C. West, we arranged to have the hitherto localised military counter-espionage services in the West transformed into Mobile Reconnaissance Commandos under the Chief of Staff West. At practically the last moment, at the end of December 1943, we received approval from the Wehrmacht Higher Command for this far-reaching proposal. Such a reorganisation had become necessary in view of the expected invasion in the West. We were nevertheless surprised when the order came through, as the all-powerful RSHA had made full preparations for the absorption of the entire Abwehr organisation into the SIPO and the SD.

The order that Wehrmacht Mobile Reconnaissance Commandos should be formed caused much gnashing of teeth in the police establishments in the West. The RSHA fought a rear-guard action in face of this accomplished fact, which culminated in an attempt to acquire a right of "professional direction" over the Commandos. This, however, left us unmoved, as we had already been saved from the fate of being swallowed up by the RSHA. The Mobile Reconnaissance Commandos remained units of the Wehrmacht, and this fact enabled us in IIIF to continue to fight our war in accordance with the principles which had guided us hitherto. My detailed instructions were to group the existing military counter-espionage authorities in Holland, Belgium and Northern France into FAK (Frontaufklärungskommando) 307, with headquarters in Brussels, and I was given this command.

These orders were as difficult to execute as they were easy to understand. Not only was there resistance to overcome from the more or less independent heads of small IIIF units, but the Commanders of Abwehr stations, who did not realise that the independence of the Abwehr had in fact come to an end, did much to sabotage this reorganisation which was removing the best of their personnel. This had, of course, been our intention. We considered that the most reliable and experienced men should be retained so as to carry on the work of military counter-espionage and not be transferred to the RSHA with the remainder of the Abwehr. We finally succeeded in April 1944 in setting up the cadres of FAKIII formations, in all respects ready for action, before the final incorporation of the rest of the Abwehr into the SIPO and SD.

Our newly acquired independence underwent a severe test in January 1944. While arresting a radio operator in The Hague who was taken by surprise communicating with a post in England, the SIPO captured one report, among others awaiting transmission, which read: " *The head of the military counter-espionage in Holland is Oberstleutnant Giskes, who was formerly in Paris and now has his headquarters at Driebergen. This man is particularly dangerous. His principal assistants are Arno and Ferdi. You are urgently warned against him. Source of this information is Oberst Stähle, Berlin-Potsdam, Hartwigstrasse 63.*

This was a severe blow. I knew Oberst Stähle personally. He had once been an Abwehr officer and had for some years past commanded the Military Orphanage in Potsdam. When in Holland he had visited me in my office in Scheveningen several times, and I had given him facilities for duty journeys into Holland ostensibly for the purpose of getting supplies for his orphanage. The SIPO at The Hague sent me a copy of this ominous message at the same time as they made their report to the RSHA. The covering letter referred unmistakably to the existence of unreliable and traitorous contacts maintained by me.

My personal acquaintanceship with Stähle was evidently known to the SIPO. In the circumstances there was nothing I could do except report the whole incident to the head of Abwehr III in Berlin, with a request that an investigation should be carried out there, since Stähle was himself in Berlin. Oberst i.G.Heinrich, the head of Abwehr III, raised the question when I was next in Berlin, and requested me to go into it with Stähle personally, no action having yet been taken. I declined, on the grounds that Stähle lived in the city, and that the local Abwehr authorities were alone competent to deal with such an investigation. Oberst Heinrich would not accept this argument, and I was only able to avoid this fatal duty on the pretext that I was due in Breslau the next day, to make a report. Heinrich finally gave me orders to see Stähle and question him within a few days of my return to Berlin, an order which I only got round by arranging with Paris for an urgent telegram to be sent to me in Breslau summoning me to a conference, with the result that I travelled direct to Paris and not by way of the capital.

The extraordinary demand made by Oberst Heinrich increased my determination to keep out of this mysterious

Stähle affair at all costs. The circumstantial background and the atmosphere of the case didn't please me at all, and I thought I could detect the existence of some sort of connection between the head of Abwehr III and Stähle, though I could not put my finger on anything concrete. I had heard in the meantime from reliable sources that Stähle belonged to an active group of the German Resistance movement in Berlin. This development acted as a further warning to me to avoid in all circumstances any mention of the names of sympathisers, which Resistance circles in Berlin were so fond of doing. Since the hair-raising Stähle incident I had more or less written off any possibility of success by these circles. I simply did not believe that those members who were known to me possessed the right conspiratorial qualifications to enable them to plan a revolutionary coup in Germany. The success of the RSHA's counter-stroke after the Attentat of 20th July was only too good a confirmation of my fears.

When I returned to Driebergen from my journeys to Berlin, Breslau and Paris I told Oberst Heinrich in confidence that it would be impossible for me to take the action demanded of me in the Stähle affair. The allegation that I had committed an unconscious act of treason would come better from a neutral authority, namely the appropriate Abwehr office in Berlin. I heard several months afterwards that the investigation had run its expected course and that Stähle had been arrested by the Gestapo on the suspicion of illegal activities. He was executed after the attempt of 20th July.

I had a somewhat similar experience in connection with another contact with the German Underground movement. At the time of the Stähle incident I dropped a personal connection which I had made for Service reasons with Oberst Freiherr von Rönne, the head of the Foreign Armies West section of the General Staff. From my experience in the Abwehr I became as doubtful of von Rönne's go-ahead methods and unbridled temperament as I was impressed with his outstanding personality and unshakeable faith in the fine ideals of the Underground movement.

The second escape of three *Nordpol* agents in December 1943 took place in circumstances almost identical with the first occasion, except that this time the men had forced their way through the ceiling of their cell and escaped over the roof. One of the men was recaptured more or less accidentally in January and brought back to Haaren. The others, agents Rietschoten and van der Giesse, remained at large, and we struck their trail only in April 1944.

In the meantime the *Nordpol* radio-links dragged on, with unimportant messages on both sides, which gave me no chance to resist the SIPO's demand that the *Nordpol* agents should be transferred to a prison or concentration camp in Germany. Up to that time my insistence on having these men always available for questioning had prevented their removal from Haaren. But now I was simply told that the greater part of them had been transferred to the prison at Assen.

In March 1944 I proposed to Berlin that we should put an end to the hollow mockery of the *Nordpol* radio-links by means of a final message. I was immediately told to submit a draft for approval to Abwehr Berlin, which must express confidence in victory. Huntemann and I set ourselves to compose a message which should fulfil not only Berlin's requirements but also our reflections on the two years' hoax which we had carried out so successfully. This message, the first to be transmitted quite openly in plain language, must not in any way fall short of the standard of the thousand-odd cypher signals which had been previously despatched. We sat at my desk and exchanged our first attempts at a suitable text in order to discover something worthy of this unique occasion. Writing rather as if we were playing "consequences", each of us composing a few sentences in turn, we finally agreed on the following:

To Messrs. Blunt, Bingham & Co., Successors Ltd., London. We understand that you have been endeavouring for some time to do business in Holland without our assistance. We regret this the more since we have acted for so long as your sole representatives in this country, to our mutual satisfaction. Nevertheless we can assure you that, should you be thinking of paying us a visit on the Continent on any extensive scale, we shall give your emissaries the same attention as we have hitherto, and a similarly warm welcome. Hoping to see you.

The names given were those of the men whom we knew to be at the head of the Netherlands section of SOE. We signalled this draft to Berlin for their approval. They were evidently occupied with more important matters, however, and we had to wait a fortnight until, after one or two reminders, we received permission to transmit the message without amendment.

I passed the plain language text to the FuB station on 31st March, with instructions to pass it to England over all the lines controlled by us, which at that time numbered ten, the next day. It had occurred to me that 1st April might be particularly apposite.

The following afternoon the FuB station reported that London had accepted the message on four lines, but had not answered calls on the other six. . . .

Operation *Nordpol* was over.

The attempt of the Allied Secret Services to gain a foothold in Holland had been delayed by two years. The establishment of armed sabotage and terror organisations, which might have disorganised the rear areas of the Atlantic Wall and crippled our defences at the critical moment of invasion, had been prevented. The penetration of the Underground movement had led to the liquidation of widely spread and boldly directed enemy Espionage Services. The complete deception of the enemy about the real state of affairs in Holland would have subjected him to the danger of a heavy defeat had he attempted to attack during 1942 or 1943. The information which we had gained about the activities and intentions of the enemy Secret Services had contributed directly to the countering of corresponding plans in other countries.

Operation *Nordpol* was no more than a drop in the ocean of blood and tears, of the suffering and destruction of the Second World War. It remains none the less a noteworthy page in the chequered and adventurous story of Secret Service, a story which is as old as humanity and as war itself.

PART THREE: BETWEEN THE LINES

UNDER the smoky arch of Brussels North station, the express from Amsterdam, which had just arrived, was disgorging a fresh flood of humanity. The voices of news-vendors echoed between mountains of luggage carried on the backs of overtired passengers, and streams of olive and field-grey figures, similarly burdened, pushed their way towards the station exits.

Groups of variously garbed workers with tired, drawn faces, a number of well-nourished war profiteers and numerous smartly dressed women with heavily painted, faded faces made up the civilian element in the stream of people passing through Brussels North in this fifth year of war. Men on leave and on duty, absentees and deserters, auxiliary and labour-corps personnel of all nationalities—from the Volga to Spain—were being pumped through the station in their tens of thou-sands by the war, the armament programmes, the defence industry and the Todt organisation.

Men from the harbours and installations of the Atlantic wall, from the bunkers of the secret-weapon sites, from airfields and workshops, depots, arsenals and camps streamed through together and dispersed in every direction, only to be followed the following day by similar masses of humanity.

This vast flood was an inexhaustible reservoir of assistance of all kinds to the plans of the enemy Secret Services, a reliable source of espionage information and a useful medium for every kind of sabotage. Since the start of the war there had been no lack of attempts by the enemy to exploit its possibilities.

The probings of the London Secret Services spread out from Unoccupied France, Spain and Switzerland towards all military objectives of importance. They struck roots in German

137

offices, on construction sites and in munitions workshops. New growths were continually appearing, at airfields and in depots, whose tendrils spread out to form a criss-cross of espionage lines. I had, when in France in 1940, already had experience of the good work of Allied espionage.

One day, in the course of the liquidation of an organisation which was being directed from Unoccupied France, some plans fell into our hands showing the French harbour of St. Nazaire, with numerous technical notations. We acted there and then and took our prize to show to the Abwehr officer on the staff of the Admiral Commanding at St. Nazaire, a worthy Fregatten-kapitän, who held up his hands in horror, simultaneously remarking that he was quite unable to prevent a recurrence of such incidents. It appeared that the plans contained precise drawings of the giant U-boat bunkers and locking installations then under construction at St. Nazaire, which included full technical details and notes of defensive measures taken from original drawings.

In the course of a few months twenty thousand foreign workers had been brought into St. Nazaire, chiefly Scandinavians, Poles and Red Spanish refugees. Their "supervision" was entrusted to half a dozen honest Secret Field Policemen from Germany, who spoke neither French nor any other of the languages concerned. A detailed report was sent to Berlin about this classic example of an impossible security situation, and there disappeared, doubtless into a heap of similar reports emanating from all the ports and military centres of the Atlantic Front from the North Cape to Biarritz.

At that time we were particularly concerned over the influx of thousands of Red Spaniards, who had crossed into France at the end of the Civil War and been put into internment camps by the French Authorities. After the French surrender in May 1940 they escaped in multitudes to Paris and into the German Occupied areas. As with so many other problems, the Occupation authority found itself more or less helpless in the face of this movement, as the following example will show.

In the course of a discussion with Admiral Canaris in the spring of 1941 a senior Abwehr officer raised this very problem. When Canaris asked for views on what should be done, it was proposed that these people should be rounded up and interned in a concentration camp (KZ). On hearing this word, Canaris put his hands to his ears as though he had not heard properly, and remarked in a friendly fashion:

"What is that—a KZ?"

"Well . . . a KZ, Herr Admiral."

At this Canaris exclaimed, his face red with anger, "I regret that I cannot follow the sense of your remarks!" With which the unhappy man's proposal died a natural death.

Since 1941, trained representatives of the Allied Secret Services had taken over the organisation and leadership of the Underground intelligence armies. French and Belgian intelligence officers set up a number of centres which were directed by fanatical resisters or well-paid adventurers. Equipment of every kind for the recording and passing of intelligence, cameras, films and radio sets, were available in plenty, and there was no lack of money, passes, identity papers, etc., for carrying on the work. We had no illusions about the difficulty of stopping this illegal activity, and we knew that the numerically inadequate German Abwehr was still only splashing about on the surface of this witches' cauldron.

There were occasional successful sweeps, it is true, by the services responsible for internal security and counter-espionage, the Secret Field Police, Wehrmacht Patrol Service, SIPO and SD, and these sometimes brought a dangerous pike out of the murky depths into the daylight. But while they broke up a few individual sections of such enemy organisations, the hydra grew new heads quicker than we could cut them off.

The situation reports passed to me by my IIIF subordinates in Belgium and Northern France since January 1944 had indicated a reasonable state of counter-espionage efficiency, but this could in no way alter the generally catastrophic nature of the situation. We were continually capturing films of detailed espionage reports, which were proof enough of the intense activity of our opponents, and it was obvious, furthermore, that we were only intercepting a fraction of the material which was getting to the enemy.

The costly operations of the Allied Secret Services in France and Belgium had undoubtedly paid them a dividend. They had decided the Underground war in their favour, despite severe setbacks, and had made preparations for the invasion which could not be controlled, much less destroyed. They had learnt how to mobilise forces in their service which could be set in motion at the moment of landing to form a formidable secret army behind the German front, an army which would appear everywhere but which could nowhere be pinned down.

The Station Square in Brussels was full of the normal midday traffic. The elegant strollers on the Boulevard Adolphe Max reminded one as little of the stress of war as did the numerous luxury shops, still showing an almost peacetime display of goods. What a striking contrast to the exhausted appearance of the Dutch cities! It was the clearest evidence of the fundamental difference of treatment accorded since 1940 to the economies of the two countries.

We had set up a temporary headquarters in the Hotel Metropole in March–April 1944. The transfer of my staff had not yet been completed, and there was a squabble with the Quartermaster's office of the local Kommandantur about the provision of a suitable office building. Brussels had the undoubted advantage of being particularly valued by the Germans as an administrative centre. This concentration of staffs and offices, some necessary, some not, had caused a situation in which even the special order from C.-in-C. West with which we were provided was unable to obtain accommodation for us. In the meantime we made do with a few rooms at the Metropole.

In one corner of the dining-room could be seen the fair hair and red face of Hauptmann Wurr. He had been three days in Brussels, with Willy, to check on a contact which had been reported as interesting by a "V"-man called "Nelis". Wurr and Willy were going to investigate.

Wurr had rung me in Driebergen the evening before to inform me that I was required urgently. He had discussed matters with his new "business friend", and the possibilities were such that I should come and speak to him without delay. Wurr was not a man to exaggerate, and I told him he could expect me in Brussels the next day. His report sounded promising.

A man who wished to get in touch with the German Abwehr had, by devious underground means, got into contact with the

"V"-man Nelis who had been given to us some weeks previously by the Lille Abwehr station. At his first meeting Wurr had heard some extraordinary things from this man. I took scarcely any trouble to conceal my doubts about the correctness of his report, and Wurr shrugged his shoulders as he replied to my objections.

"I have started an investigation," he said, "as to the correctness of the information which he has supplied about himself. He calls himself Christiaan Lindemans, and says he is a Dutch civil servant with a house in Rotterdam. He came yesterday to the IIIF town office in the Jardin Botanique in Brussels, having been brought there by Nelis, and spoke up with a remarkable display of confidence. The man is either a quite genuine mine of information or else the most dangerous character we have encountered so far.

"I evaded his first question—whether he was addressing the head of the German counter-espionage, whereupon he remarked somewhat abruptly that he would not stand for any tricks. He would deal only with the Chief, and anything less would be useless and a waste of time. When I asked him how he could illustrate his importance from our point of view he named a number of individuals and links which had in fact been recorded in our Top Secret files. For instance, he claimed to have contacts with the CSVI group in Amsterdam and with the financial supporters of the Underground movements in Brussels and Paris. In this connection he named the local and the Paris representatives of the firm of Philips. He will be at the Jardin Botanique again at 1800 today, and I have told him that he can then speak to the Chief."

"What does the man look like?" I asked.

"He is a giant of a fellow, who gives an impression varying between extreme brutality and harmless simplicity. Nelis declares that he is one of the most active and sinister figures in the Underground movement in the West, who has a record of bloody affrays with the German police and who shoots on the slightest provocation. His cover-name in Underground circles is 'King Kong'. And he does in fact give the impression of a gorilla. I shouldn't like to come up against him by myself. He is a double-dyed and dangerous menace, on whom it is impossible to rely!"

"Did you say his name is Christiaan? We'll double it and call him 'CC' from now on. Who else will be at the Jardin Botanique?"

"Willy and Nelis will be there, besides you and me. It'll be better if we are not disturbed by the other IIIF people who live there."

"Fix it so that you are there to meet him. I will arrive a little after 1800 and it will do him no harm to wait for me a quarter of an hour. Well—until then. I'm going to the Abwehr office and you can reach me through Major Göhring."

When I entered the house in the Jardin Botanique, somewhat late, at nearly 1830, my three assistants were seated with the stranger, whom I had that afternoon rechristened CC, round a large table. CC towered a full head above all of us as they rose on my arrival. Wurr had not exaggerated—there was standing before me an athlete with the head of a child. He was evidently about thirty years old.

"Gerhardts," I said, introducing myself, "please sit down."

We sat down, all five of us, and CC and I regarded one another closely and unceremoniously.

"May I ask you to explain what brings you here?" I started the conversation. "I have heard it said that you have contacts with the Allied Secret Services, and I shall be grateful if you will tell me in brief terms who you are, what you want from us, and what you have to offer."

CC replied in fluent German. "If I am not mistaken," he began, "I am speaking to the head of the German counter-espionage. I wish to address my proposal to him alone, as I do not expect to get satisfaction from anyone else. My personal particulars as given yesterday to Herr Walter (Wurr) are genuine. I am Christiaan Lindemans of Rotterdam, and I have worked for the English Secret Service since the spring of 1940. For the last six months I have brought in my youngest brother to assist in getting English airmen out of the country. He has been discovered, arrested by the SIPO, and is now under sentence of death pronounced by a German military court. I feel myself responsible for my brother's fate, since it was I who introduced him to this work. If you can arrange to have my brother freed, I am ready to hand over the whole of my knowledge of the Allied Secret Services. I know the 'Underground' from the North Sea to the Spanish frontier. You may assume that after five years of work for the Allies I have experience and contacts which will be of great value to you. But I must specify one condition. I am aware of the methods which the German military counter-espionage is accustomed to use, in contrast with those of the Secret Field Police and the SIPO. That is why I have come to you and not to the police. I could not imagine that you would proceed to make mass arrests of all my friends. But that is not the main point. The decisive factor is whether you can give me your word that my brother Henk Lindemans will be set free. The rest I leave in your hands."

He had laid his heavy fists in front of him on the table like a pair of blacksmith's hammers, and I noticed that he moved his right arm, which was evidently injured, by means of his left. A pair of steel-blue eyes gazed at me in strained expectation beneath a broad, low, heavy forehead. Wurr, Willy and Nelis were listening silently to our conversation.

"I do not know your brother's case. If he has been sentenced simply for the crime of helping prisoners to escape I think I can promise you that we can have him set free. I will go into the matter at once. If you can satisfy me that you are playing us no tricks, and if the information which you give us proves that you are the man you make yourself out to be, you may rely on it that your brother will be free in a week, at the latest. You are, by the way, quite correct in assuming that the Military Abwehr is not interested in mass-persecution. Our main problem is to watch or to liquidate leading individuals, or to control such persons, in order to find out the intentions of the London Secret Services. Future decisions depend on what you are prepared to tell us."

CC had held me with his eyes and listened intently to what I had to tell him. It was easy to see that the promise to release his brother had been decisive. He rose abruptly, as though unable any longer to control his agitation. We watched him with astonishment as he strode up and down the room, opening his heart more to himself than to us.

"For the past five years I have been impelled by a single thought—to do my utmost for the Allied Secret Service, without thought of thanks or reward. I have been met with ingratitude, mistrust and betrayal. If you only knew how many weaklings, place-seekers and collaborators, who have used their connections with the Germans simply to enrich themselves, are now starting to come over to us because they believe that the defeat of Germany is imminent. If you knew this you would understand me better and would realise why I have come to you. The men through whom we carried on the Resistance during the first years of the Occupation have nearly all gone—dead, arrested or just disappeared. Of the remainder, there are only a few whom I can trust. Leave them in peace! I will guarantee that in due course you will learn a great deal about the plans of the Underground and of London. Hand me over my brother and then make use of me as seems best to you. King Kong, as they call me, is friend or foe. I should like to be your friend from today. I have often heard, and always believed, that those who work conscientiously for you are treated properly. As regards confidence, I have had enough experience to know that there is no such thing as half-confidence. That goes for me too—all or nothing. I shall show you that King Kong can be relied on, and what it means to have him for a friend."

The man had spoken ever faster and more feverishly. He went suddenly to one side, grasped a thick briefcase and emptied its contents on to the table.

"Have a look at this. You'll find something that will interest you."

After CC's impassioned speech we couldn't help laughing as we looked at the pile before us. A heap of papers, forms, documents, parcels, personal identity papers, a number of German official rubber stamps, facsimile signatures, a 9-m/m. pistol and several thick wads of French, Belgian and Dutch bank-notes lay on the table. The possession of any one of these forbidden articles would be cause enough for an inhabitant of the Occupied areas to be sent to a concentration camp by the police. CC observed the surprise of his former opponents with pride, who seemed ready to recognise this curious collection as proof of his underground work. Together with stamps, facsimiles and pro-formas, the travel documents for the OT and Wehrmacht had Field Post numbers and signatures already inserted, and lacked simply the name of the user.

I selected from among them a few particularly good

"museum" pieces to be passed on to the Security sections responsible for the organisations concerned. CC grinned when I asked him to let me have them to use at my discretion. There was no need to reassure him that the source would be kept secret from third parties.

After Wurr had made an impression of each of the rubber stamps, etc., and a note of the Field Post numbers which had been used, CC packed the considerable quantity remaining back in his briefcase.

"I don't have to give you advice," I said, "but I suggest that you be very careful with these things. I shall have more important work for you in future than playing with such toys."

In the meantime it had occurred to me that it would be preferable not to let Nelis see all the cards in my hand, apart from all those which I expected CC would have in his. We knew too little about him, and I wished to cut him out of the affair as soon as possible. The services which he had previously rendered to Abwehr Lille were no reason why I should let him in on the exploitation of the new contact which he had produced. I therefore arranged for CC's main report to be given to Wurr and Willy the next day. For this evening I contented myself with general discussions, in which CC gave of his best from the depths of his manifold experience. He left us about nine, making the excuse of an urgent meeting.

Three days afterwards Wurr arrived in Driebergen with the collated report from CC. They had been working on it day and night, and the results amounted to a convincing picture of every kind of clandestine activity in the Western Occupied areas. Dozens of important courier-lines, complete with their safe-houses, methods of contact, pass-words and frontier crossing-points, several hundred names and cover-names of paymasters, pay offices, couriers and agencies of all kinds belonging to various secret organisations formed a mosaic of seditious activity such as we had never before been able to acquire through penetration from the outside. The check made on Lindemans' personal details had confirmed the authenticity of this information.

The measures to set free Henk Lindemans, who was due to leave for Germany with the Labour Corps, were now set in motion. Nelis was cut out of the case, and employed from that time on by Major Wiesekötter in Holland—the risk of his maintaining contact with Lindemans could be run, seeing that he was well-known in Belgium and Northern France as an Abwehr informer. I detailed Wurr to be the single link with Lindemans, and Willy to be contact man. The material which CC had handed over was passed to the appropriate Abwehr departments for information only, the nature of the source being concealed, with the stipulation that in no circumstances should action be taken against any person named without the prior approval of FAK307. My safeguards succeeded, in that in only two cases was action taken against small groups which were among those named by Lindemans.

In the meantime Wurr was carefully initiating CC into his new role, and I took part in the discussions from time to time in order to increase his confidence. This was no longer necessary when his brother Henk was set free shortly afterwards. Our plan was for Lindemans to play the part of a leading member of the Underground, who knew everything and had a hand in everything, but who wished to remain in the background, in order to be available for the bigger tasks which would be ordered by the Allied Secret Service at the moment of invasion.

The discovery of the time and place of the Allied landings had become the supreme objective of the German Secret Services in the West, and everything else was subordinated to this end. It was evidently considered that a successful landing would be irreparable, unless it could be at once thrown back into the sea by superior forces. The heavy preponderance of the Allied Air Force could reduce the German counter-attack to impotence if reserves had to be brought all the way from France; and the German position would become even more serious if a mistake should be made over the invasion point and big troop movements become necessary on a line parallel to the actual front. The staffs of the French and Belgian Secret Services and the well-organised sabotage groups, which were all controlled from London, were waiting the transmission of agreed code-words over the BBC before going into action, either on or after Day X, the date of the Invasion; the attempt to decypher the dozens of BBC messages which were passed daily thus became the main problem of the German counter-espionage.

We were hoping that we could introduce Lindemans into sufficiently close contact with the operational headquarters of the enemy Secret Services to give us timely information of the time and place of Day X. After a short while Lindemans was sent into Belgium as a representative of the Dutch Underground, to hurry up the supply of arms from Belgium into Holland, promising to get the Dutch, who were very short, large supplies by this means. Towards the end of March Lindemans reported to the effect that two English agents who were working in Holland were to be picked up from the coast of Zeeland by MTB or submarine, an operation arranged over one of the radio-links operating from that country. The radio situation had already deteriorated from our point of view before the end of Operation *Nordpol*. The FuB service was watching at least six of these stations in Holland without yet being in a position to liquidate them, and captures made by the SIPO were soon replaced by similar transmitters in other areas. It was evident that London was now profiting by the experience which it had bought so dearly in *Nordpol*.

In order to get on the trail of the agents he had reported, we told CC to propose to the courier group that they should give him the job of bringing the two men to Zeeland. Permission was required for journeys to Zeeland using normal means of transport, and the travellers needed special passes to get them on to the island. Lindemans offered to procure these, and was accordingly given the photos of the men and other particulars. We were not particularly surprised to discover that these two were no other than Rietschoten and van der Giessen, who had escaped from Haaren in December, and whom we had supposed long since safe in a neutral country or in England. What was to be done? We were not particularly inclined to let the enemy's plan go through. These two men, however, could no longer do us any direct harm by what they could tell in England, and the influence and standing of Lindemans, which was my immediate concern, would be immensely increased if he were able to carry the business through to a successful conclusion.

We decided to let Rietschoten and van der Giessen be picked up undisturbed, and arranged for the necessary passes to be provided. Two days before the date of departure Lindemans reported that these two had been instructed to take with them to England a whole consignment of Leica films. These represented a collection of important espionage reports received over the previous few months, which required rapid transmission to England. The seriousness of this new information wrecked our plans completely, as we could not possibly stand aside and watch this material pass over to the enemy.

We were unaware whether any coastal patrols off the mouths of the Maas or Scheldt, or off the Zeeland islands, figured in the enemy's invasion plans, although there were indications in the possession of C.-in-C. West which appeared to support this. Our particular interest was whether an MTB or submarine would be used for the pick-up. In order to get hold of the espionage reports there was nothing to be done except to rearrest Rietschoten and van der Giessen. This was effected by means of a pass-control on the way to Bergen op Zoom. The passes of the two men were ostensibly recognised as having been forged, and this served as a pretext for their arrest. In this way we got hold of about thirty Leica films, which, when developed, showed very clearly why the enemy was prepared to send a warship to collect them. In them a complete picture of the enemy's intelligence targets lay before us. We were accustomed to come across good work in the way such reports were prepared—the training in England saw to that—but in this case specialists had worked with an accuracy and efficiency which we had not previously encountered. Particularly worthy of note was an almost complete picture of the secret-weapon sites in Holland.

This latest feat of the agents of the Dutch headquarters in London—the Bureau Inlichtingen, known for short as the BI—increased our already considerable respect for this adversary. In 1942 and 1943 the BI had had to overcome serious setbacks. Many of their agents and transmitters had fallen into German hands, but we had not succeeded in playing back a single one of these radio sets. A severe blow had been struck against a group of BI radio operators working from Groningen, Drenthe and Amsterdam, directed by a teacher from the Navigation School at Delfzijl, and we had not heard much of them since June–July 1943. Further expansion of this organisation had, however, by no means been stopped. Drops of agents continued, evidently without using Reception Committees, and we heard about them through other channels. Major Wiesekötter was in charge of operations against the BI, and I will describe one or two of the more characteristic incidents which took place.

In the middle of August 1943 the ferry-boat plying between Enkhuyzen and Stavoren picked up a rubber boat with ten occupants in the early morning hours—the crew of a bomber which had been shot down during the night over the Ijsselmeer. Two of these men disappeared when the ferry got to land, while the German harbour police placed the remaining eight under arrest. It was established that this was the crew of a "specialist" aircraft. Apart from the eight men of the crew, there had been three agents on board who were to be dropped into Overijssel. Two of them had escaped, and the third had apparently gone down in the sunken bomber, either dead or seriously wounded. The airmen were clearly in ignorance of the tasks given to the agents, but a search of the aircraft might give us a clue. The Luftwaffe headquarters in Amsterdam smiled at our insistence on searching for the wreck of the aircraft, their experts maintaining that no one knew its position, quite apart from the fact that over a hundred similar wrecks lay at the bottom of the Ijsselmeer, which made it out of the question to find and to raise the one we wanted. Nevertheless, encouraged by the promise of a substantial reward, a number of naval and fishing craft did go to the spot and search, with the result that the sunken aircraft was brought on shore at Schellingwoude a week later. The body of the third agent was found, together with papers which clearly indicated that the BI was involved. A quantity of bank-notes and a photograph of Colonel Somer, the head of the BI, which was evidently intended for use as a pass, completed our haul.

In another case, which took place in the autumn of 1943, the situation reports of the Luftwaffe Group headquarters which issued details of "specialist" flights had caused suspicion to fall on to an area between Vught and Grave as a possible dropping point. We here enlisted the services of the motor patrols belonging to the local German Area Commander, which succeeded in picking up a suspicious box, after a drop had taken place. Among other articles this contained a very worn civilian suit, which, according to the tab, had been made some years previously by a tailor in Eindhoven for a certain Herr Shoemakers. We could scarcely credit that any man would let himself be dropped as a spy with genuine tailor's tabs as part of his luggage, but we made some enquiries about this Herr Shoemakers. Results were rather long in coming, but when they did they astonished us! Herr Shoemakers was a member of the technical managerial side of the Philips factory in Eindhoven, had been there just previously, and had left in the meantime for Stockholm with, according to our information, the approval of a German officer in Berlin. That was quite sufficient for me to become interested. The office in Berlin which had sponsored Shoemakers' journey was asked to arrange for him to proceed to Copenhagen for technical discussions, and I intended to be there myself when this gentleman arrived so as to have a closer look at him. I accordingly wrote to my IIIF colleague in Copenhagen, Oberleutnant F——, but on the evening before my departure he telegraphed to say that the meeting could not be arranged, as a general strike which had just broken out in Denmark had cut all communications. From then on the Shoemakers case was handled by AstIIIF Copenhagen, without my hearing anything further other than a remark made in Berlin to the effect that Shoemakers was thought to be an agent of the BI. It is my impression that Shoemakers had a highly placed protector on the German side with whom Oberleutnant F—— was in some way connected. F——, incidentally, was arrested and shot after 20th July.

Between the autumn of 1943 and the spring of 1944 the BI must have succeeded in setting up a highly efficient espionage organisation, possessing good courier lines and numerous radio-links between Holland and England. Although much useful material fell into our hands through our controls inside the courier lines in Belgium and France, we never succeeded in destroying or even in seriously weakening this organisation. The documents which we found on Rietschoten and van der Giessen were sufficient proof of that! The SIPO started a wild hunt for the men behind this affair, and for the copies of the films which had almost certainly been made. We heard little of any results, which was usually a sign that the police had not got very far in their investigations.

Rietschoten and van der Giessen were taken back to Haaren after their arrest. On the way they once again attempted to escape, and despite their handcuffs they succeeded in pitching their SIPO escort out of the moving car. They did not get far, however, and were placed under strict guard when they reached Haaren. Sometime afterwards Major Wiesekötter received a laconic report from the SIPO to the effect that these two brave and determined men had been shot while attempting once again to escape.

At the beginning of April a meeting was arranged with Lindemans at Driebergen. I was then living in a house outside the town, all by itself in the pine woods. A car collected Lindemans at Amersfoort and brought him by a devious route over the heath to my back entrance. It was a few days after the arrest of Rietschoten and van der Giessen, and Willy had reported that CC wished urgently to see me.

We shook hands as usual and sat down at the table, Wurr to my left and Willy to my right. CC appeared uneasy, incoherent and morose, and I did not like his whole attitude. He declined both cigarettes and brandy. I was even less happy when he drew the 9-mm. Colt, which Willy had given him some time previously, again and again from his inside pocket, loaded and unloaded the magazine and clicked the safety-catch up and down. Placing his elbows on the table and regarding me with a mixed expression of spite and stupidity, he was behaving like a child trying to test the patience of a grown-up. We knew him to be "trigger-happy" and armed to the teeth, and we had heard that he had actually used his gun on several occasions. Supposing he had it in mind to shoot up the head of the German counter-espionage and his two assistants?

I concealed my disquiet and tried to show him that I was up to all his tricks. Walking quietly across to my desk, I picked up a heavy American army 12-mm. Colt and asked CC, as I handed it to him, whether he had ever come across one of these. He put his gun away and examined the new one while I explained its method of working.

"If you like, Willy will get you one," I said, as I took it back from him.

During our long conversation the gun lay, apparently neglected but ready for use, beside my teacup, and CC stopped playing with his own pistol. When I rang the bell an hour later to give the driver his return instructions CC was once again his old self.

We had discussed a plan which was to bring a consignment of weapons and a radio set into our hands. Two days previously Group CSVI had received a drop from a "specialist". CC brought the material over on the following morning, in a car provided by us, from its temporary place of storage. Shortly afterwards the SIPO raided the CSVI headquarters in Amsterdam in co-operation with the FuB section of the ORPO and captured the radio set, which had not yet been put into operation, with all its associated material intact. It seemed that an opportunity had presented itself for playing this set back. Knowing that CC had had a hand in the affair, Hauptmann Kienhardt of the FuB section passed the set over to me. This obvious piece of initiative drew the wrath of the SIPO on to him, and I once more had the pleasure, after a long interval, of a ring from Schreieder. The head of IVE was extremely angry. The threats of what would happen if I refused to hand the set over to the SIPO tumbled over one another. If I wanted to press my claims to the set I should have no option but to publicise all the preparatory work which had been done by CC in the CSVI case. The set was simply not worth the price, so when Schreieder came on the line again half an hour later I told him that he could collect it from Driebergen when he wished. This was my last conversation with Herr Schreieder during the war.

The play-back of this set which was subsequently started by the SIPO soon led to the receipt of messages from London which indicated, most politely, that the team in Holland should give up their attempts to play this game! We heard about this comedy when the specialist from SIPO IVE who was running the play-back asked the advice of Huntemann, my English language expert. He produced a thoroughly ambiguous English text which had several possible alternative meanings. Huntemann could in all honesty only come to the same opinion as the English and advise the SIPO officials not to go on wasting their time.

This experience only proved afresh that our former fruitful co-operation with the SIPO was finally at an end. It had deteriorated to a gulf of mutual mistrust and antipathy which simply could not be bridged. We now had to face an additional front, and we well knew how dangerous our new opponent was. When I moved to Brussels with my staff in May 1944 I left Major Wiesekötter, in his capacity as head of IIIF in Holland, with a very complicated situation to deal with. He could no longer rely on any help from the SIPO, and had simply to make the best of a bad job.

In January 1944, in addition to the IIIF's sections in Holland, Belgium and Northern France, Berlin placed Station P in Holland under the command of FAK307. As previously mentioned, Kapitän Patzig, the Commanding Officer, had been in charge of IIIF contacts in Holland before the war. This state of affairs had been brought to an end by the events of May 1940 and could not be resumed. I had got to know Patzig personally in the autumn of 1941 and we had agreed to help one another in IIIF matters. He had, however, had little to give me, and the mutual interchange had gradually come to an end. Abwehr headquarters had given him as well as myself the task of trying to discover the German source of the report which Major Sass, the last Dutch Military Attaché in Berlin, had telephoned to The Hague on the evening of 9th May 1940. Major Sass had reported that the "delivery" was expected in the early hours of 10th May. Berlin considered that it would be possible to discover the source of this information from material available in Holland. My office did not succeed in fulfilling this task, but the investigation served to contradict even more the evidence against General Oster, who was under suspicion of having disclosed that the attack against the West was imminent. This circumstance served to explain the special interest shown by Abwehr Berlin in pursuing the matter into Holland.

I do not know whether Patzig was any more successful in this than I was. It was unfortunate that 1944 had not fulfilled my earlier hopes of assistance from Station P.

When our headquarters was set up in May in a large and secure building in the Place de l'Industrie in Brussels, mutual relationships in Belgium had already become uneasy and unsafe. At that time no additional protection was necessary for staff headquarters in Holland. In many parts of Belgium, on the other hand, conditions already verged on civil war and necessitated special precautions. Apart from CC, FAK307 had no direct contact with the Underground, and it was thus easier for it than for the detachments stationed in Brussels, Ghent, Lüttich, Antwerp and Lille to remain concealed and so avoid attacks. FAK307 had the duty of co-ordinating the activities of all the military reconnaissance detachments, evaluating their reports, and providing the link with the No. 3 headquarters in Paris, which collated the results of the German counter-espionage in the West. The situation reports which we signalled daily to No. 3 headquarters, West, indicated a steadily deteriorating state of affairs. The number of ambushes, attacks, and incidents involving the use of explosives by the Belgian and Northern French Underground increased slowly but steadily in April and May, with a big jump after the start of the invasion. In June, July and August we had to report an average of thirty to forty armed incidents and sabotage attempts daily. The principal objectives were railway junctions, bridges, locks and every other sort of traffic and communication centre. At the same time a bloody terror broke out against everyone suspected of friendly relations with the Germans or who had joined up with them in the Belgian SS formations. The wives and children of Belgian SS men who were fighting on the Russian Front were in many instances brutally murdered during that period. The split caused by the Hitlerite ideology within the population increased to a yawning gap as a result of the reprisals which were threatened. Bullets were freely used. Anyone who was known to be engaged in Intelligence work against the enemy or to be working against his agents could expect to receive a salvo of machine-pistol bullets round every corner. Bloody affrays in which Abwehr officers and GFP officials were shot to bits increased in number and forced us to take additional security measures.

CC himself suffered from this increasing tension. One day a frightened SIPO official shot him.

The SIPO had received information to the effect that a well-known member of the Underground would be visiting the office of the Currency Control Service in Rotterdam at a certain time. No one appreciated that this was actually CC. When the arrest squad appeared, Lindemans had shown no haste in obeying the command "Hands up"—I could not establish whether he had tried to draw his own gun—and was taken off to hospital with a bullet in his thigh. Fortunately the wound did not seem to be dangerous.

We then organised an "abduction", which went off according to plan, and it was inferred that his friends in the Underground had liberated him. He was up and about again within a fortnight, having been nursed back to health meanwhile by one of his many lady friends. Generally speaking, women were Lindemans' weak spot, and it was the only sphere of activity in which he had to be warned to be careful. I was the more astonished, therefore, when he arrived one day, very cast-down, to ask me if his wife could be set free, as she had been arrested by the SIPO, together with their child, during a police sweep in Paris. They had been staying in the Hotel Montholon and had been caught up in the process while this haunt of the Underground was being cleaned up. A few days later I was able to tell him that we had been able to set his wife and child free, of whose existence I only learnt for the first time through this affair.

The state of communications at that time made a journey to Paris an undertaking whose duration could not be forecast. CC, however, could not be dissuaded from going there, so as to reassure himself of the well-being of his family. The railway line between Paris and Brussels was almost continuously out of action from sabotage and air attack, and many trains had been diverted via Lorraine and Luxembourg.

I heard to my relief a week later that CC had arrived back in Brussels. The man was capable of the most improbable surprises. You could always remember when you had last seen him, but it was quite impossible to rely on when or whether one would see him again.

In our military counter-espionage activity covering all possible sources of intelligence concerning the impending Allied invasion the IIIF out-station in Antwerp had come across a potentially useful line. This station had made excellent contacts with underground groups in Holland which were concerned with the collection and concealment of shot-down

Allied airmen and which organised their onward journey thence into Belgium. Contact men from the Antwerp IIIF office, purporting to be members of the Belgian Underground, would meet the airmen on the Dutch-Belgian frontier and bring them to Antwerp. Here a small number of language experts from the Abwehr staff occupied a large house which was known as the "Antwerp airmen's pipe-line" and acted the part of a secret "central committee" of the Belgian conducting organisation.

On their arrival the airmen had to fill up a printed form in English, ostensibly in order to confirm their identity to London by radio. It was explained to them that this full description was necessary for protection against German informers who had tried to penetrate the organisation in the guise of Allied airmen. If any one of the escaping men should not be accepted by London as genuine he would come at once under suspicion of being an informer and could not then expect to leave the house alive. There was no doubt therefore about the care with which these questionnaires were completed, and they provided good leads for subsequent serious interrogations. After a stay of two or three days in the house, the airmen were told that they had been approved by London and that their onward journey to the French frontier would start the next day. As there were on occasions not more than two or three airmen in the house, a close relationship soon sprang up between them and the gentlemen of the "Belgian committee", and this relationship was cemented by good eating and better drinking in the company of young women "friends of the house". Interrogating officers from the Luftwaffe appeared, disguised as Belgians and "helpers of the committee", and the results of their conversations produced important intelligence about all aspects of the Allied war effort.

Transport for the men was provided in the form of a car with Belgian number-plates driven by a IIIF man in plain clothes. A patrol of the GFP would stop the car on the Brussels autobahn and examine the papers of the occupants, who would be placed under arrest until their identity could be established. The airmen's speech naturally failed to correspond with the description of "Belgians" on the documents supplied by the "committee", and they had then once again to supply accurate particulars of themselves so as not to come under suspicion of being spies. They were eventually taken over as prisoners of war by a Transport Group of the Luftwaffe.

Over one hundred and eighty Allied airmen passed through the Antwerp pipe-line between March 1944 and the evacuation of the city in September. At this period the results of interrogations carried out by the methods described provided one of the best sources of the German military intelligence.

The riddle of the place and time of the invasion had become the nightmare of the German military leaders, and it was intensified when the number of code messages passed daily over the BBC increased to nearly two hundred in the last week in May. Although the German counter-espionage was in fact able to work out the sense of part of these messages, it did not succeed in obtaining any definite indications about the Allied invasion programme. The increase in the number of transmissions had simply to be taken as a signal that the invasion itself was imminent.

After the great attack we had no longer any need to worry our heads over these BBC messages. Through the events which followed we were well aware that the Allied secret armies had gone into action behind our backs. Within a short time our maps were showing large parts of Belgium and Northern France as being known partisan areas. The passage of Wehrmacht vehicles through these areas was only possible in company. Single cars were shot to bits. Our contacts with CC failed to provide us with any information of importance concerning local enemy activity.

News reached us on the morning of 21st July 1944 of the attempt on Hitler's life and the revolutionary coup in Berlin. The uprising had led to sympathetic action in Paris, where, on the receipt of the news about the events in Berlin, the C.-in-C. France had had the quarters of the SIPO and SD in the Avenue Foch surrounded and the entire personnel of these two departments taken away to Fort Vincennes. It seems that they were to have been shot out of hand. A delay in the execution of this order was caused by the obscurity of the situation in Berlin, and it was revoked as soon as the failure of the Berlin coup had been established.

The messages concerning these events which arrived in Brussels hardly gave a clear picture of the situation. It was confirmed on the morning of 21st July that the attempt by the Resistance movement had failed, but nobody knew what was coming next. We had no illusions that the RSHA's counter-stroke would confine itself to those directly implicated in the operation and to their friends who had provided assistance of various kinds. With the precedent of 30th June 1934 in mind, we looked forward with some misgiving to a general purge, regarding which the bloodcurdling announcements from Berlin left no room for doubt. If there were ever to be any relaxation it would be only because the discontented masses could no longer be kept silent.

The Attentat which missed fire had one important and unforeseen consequence. It stimulated men's deadened spirits and blew the dull fire of discontent into a blaze. The Nazi Commissars among the officers were no longer able to suppress discussion among the men about the necessity and the possibilities of continuing the war.

I received a report from Major Wiesekötter at the beginning of August which indicated for the first time that there was a crisis in the Abwehr position in Holland. In London MID-SOE had been replaced by the newly constituted BBO and its efforts had this time been crowned with success. The organisation of sabotage and resistance groups seemed to be developing, and the supply of weapons and material by air was making rapid progress. They had not yet been ordered to go into action, but this would not be delayed any longer should the Allies push forward into Belgium. Collaboration with the SIPO had since 20th July ceased absolutely. Even our old contacts with individual experts who did their work objectively and with a trained sense of duty were now useless. One could sense that instructions had been given in high quarters that the military counter-espionage was to be isolated.

Wiesekötter finally asked me to help. Help? How and where? An attempt must nevertheless be made to change the situation in Holland. So I rang up Wiesekötter and asked him to arrange for me to see Naumann, the head of the SIPO. If I were to penetrate the lion's den I could at least expect to have the situation clarified. Something had to be done.

On 9th August Wiesekötter informed me that my interview
with Naumann had been fixed for 1100 on the 10th. I asked
him to accompany me, and we arranged to meet at The Hague
the following morning.

The hot midday sun hung over the Plein as I parked my
car shortly before eleven, after a fast drive from Brussels.
Exactly three years previously I had first arrived at The
Hague to take up my duties, and it now seemed my time must
be nearly up. The sentry at the main entrance of the SIPO
headquarters buildings saluted as usual, and Major Wiesekötter and I walked upstairs to the office of the chief on the
second floor. We were evidently expected. As soon as the
Sturmführer in the outer office had announced us we were
shown into the room, where Naumann and the head of Section
III, Sturmbannführer Deppner, were standing waiting for us.

Our greeting was limited, on my side to "Good morning,
meine Herren" and a bow, which was returned. Deppner
waved us towards two chairs in a corner of the large room and
planted his Don Quixote-like figure beside the desk, behind
which his chief sat down. The atmosphere was noticeably
chilly. I had not seen Naumann since the last drop in October
1943. Neither of us had sought an opportunity of settling the
serious differences which had made their appearance during
the last phase of *Nordpol*, and in connection with the formation
of the Frontaufklärungskommandos. Since that time Germany,
the leaking boat in which we sat, had been brought to a
sinking condition, and a shipwreck was impending, from which
the will of Hitler and his followers would ensure that there
would be no survivors. We were at this moment sitting
opposite two of the instruments of this will, and there was that
in the air which impelled me to the utmost caution. It had
become clear as daylight that not the slightest weakness must
be shown.

We regarded the fanatical Deppner as a blind tool in his
master's hand, his predecessor, Wolf, having been by comparison a harmless type. Naumann's robust figure seemed to
have become even broader since the previous year. There were
malicious creases in the rolls of fat on his round face as he
opened the conversation by enquiring as to what he might
owe the honour of this visit.

"Major Wiesekötter has reported to me," I began, "that for a number of weeks past he has received no reports or information from your stations, and that his officers have been put off by your officials with various excuses. May I request an explanation of this conduct?"

Naumann stood up, his hands crossed behind his back, and advanced towards us with slow, heavy steps. An image sprang to my mind, of a hyena creeping upon its prey.

Stopping at arm's length from me, he thrust his face forward and burst out: "There are no longer sufficient grounds for making reports to the Abwehr, by which I mean you—even though you may call yourself 'Frontaufklärung'. You know, I suppose, what has been occurring in Paris and Berlin? Or would you have me believe you have not yet heard about all your friends having been arrested?"

This had almost the air of an interrogation! A proper "shock-question"! I forced myself to remain calm and indifferent.

"I haven't had much of an opportunity to form a clear picture of recent events. All that I hear from Berlin is contradictory and I have to rely on the newspapers for reliable information. . . ."

Naumann broke in rudely—my opinion of the Goebbels Press had upset him: "Please don't imagine you can still deceive us! We are fully informed about you, and I am in agreement with my colleagues in Paris and Brussels that you will have to disappear as soon as possible."

This was a threat—clear and unmistakable, and it was time to cut the conversation short. I rose with a jerk and turned to Wiesekötter.

"I have nothing more to say, and I am sure you will agree that there is no point in continuing this discussion."

We moved towards the door—Wiesekötter like an automaton beside me. Our brief bow remained unanswered, and the surprise of the blustering Naumann at the unexpected result of his threats was reflected in the face of his angular adjutant. We crossed the outer office in three quick steps and hastened downstairs.

"Have you a gun handy?" was the first thing I whispered to Wiesekötter. "Perhaps he was only bluffing, but if those scoundrels really believe that they can get something on us we shan't get past the sentries at the bottom. I've no wish to disappear into some hole or other and I'll shoot before they arrest me!"

Wiesekötter was pale, and it was plain that he hadn't fully grasped the situation. The last flight of stairs lay before us and there was no one to be seen. The door of the guardroom was standing wide open and the telephone bell shrilling as we passed the saluting sentries and stepped out into the sunlight.

The faithful face of my old driver Ramm looked at me questioningly as he opened the car door. "Driebergen, Ramm. We're in a hurry!"

The Beukenstein Park lay in its midday sleep. As we entered the building the NCO on watch reported: "Nothing particular to report."

Had it been an empty threat—that I should "have to disappear"? I didn't trust the seeming peace of things. Dozens of Wehrmacht officers had disappeared in the past few weeks, whose fate was unknown.

Half an hour later we were on our way to friends in Germany. I rang up Brussels and Driebergen from the Abwehr office in Köln and spoke to Wurr and Wiesekötter.

"Anything particular?"

"No, Herr Oberstleutnant."

"Thank you, that's all."

Before driving back to Brussels next evening I enquired once more about the situation there by telephone from Köln. All clear! During the night Wurr and I took some special precautions at the Place de l'Industrie. If anyone had designs on us they would have to plan well ahead.... Our precautions, taken under the pretext of protecting ourselves against attacks by the Belgian Underground, showed the extent of our experience of the methods used by the SIPO and SD. Meanwhile events piled one upon another, and there were good grounds for believing that other worries had become more urgent for the SIPO chiefs in the West than a search for an excuse to remove the unpopular head of the Frontaufklärung.

During this period, in France the dam broke which had previously held up the flood of Allied divisions streaming without a pause across the Channel. Enemy armoured spearheads thrust north and east across France. Paris fell. The fronts were continually in movement, and it was impossible to say where they would be brought to a stand. Our radio-link with the German headquarters in France was interrupted from time to time as it moved north, and by the end of August it was no longer possible to obtain either a clear picture of the situation or any forecast of future developments. We waited in vain for detailed instructions covering the possibility that Belgium would have to be evacuated. This picture of impending catastrophe was heightened by hysterical cries from Berlin calling on every man to stand, fight and die on the spot in which he found himself. If it were possible, they only increased the confusion, since every commander, regardless of the size of his command, now issued instructions on his own initiative for "all-round defence", which were nevertheless changed into orders to retreat at the last moment, usually when it was already too late. No one had dared to give orders in good time for the enormous depots and magazines to be evacuated—their irreplaceable stores fell into the hands of the rapidly advancing enemy, when they were not looted by the local inhabitants.

We continued our work in Belgium amid growing difficulties, but untouched by the tumult in France. Our very mobile small units were in little danger of being cut off unawares or of having to break away and leave their equipment behind. On 25th August King Kong brought in a report which purported to emanate from the head of the *Armée Blanche*. The report indicated that the main thrust of the Allies was directed at the Dinant area, with the intention of advancing via Namur in the direction of Eindhoven so as to seize the river crossings at Nijmegen and Arnhem. The subsequent attack would follow from a bridgehead thrown across the Rhine and Waal, down the Ijssel and towards the German North Sea coast.

On that day we were able to pass this message over the radio-link which had been re-established with No. III headquarters West, which had now moved back to the Luxembourg area. An attempt at confirmation of the report was unsuccessful, but the actual development of the Allied attack during the next three weeks established the correctness of the information.

We had several local successes in these weeks which are scarcely worthy of mention, apart from a short play-back of a radio set taken from an agent who fell into our hands in the Lille area. The night before the occupation of the city we accepted a last drop of weapons, which was controlled by the arrested agent in communication with the supplying aircraft. Huntemann, who was handling the agent, arrived in Brussels a few hours afterwards and reported that the operation had been successful. When I enquired about the present whereabouts of the agent he indicated by a movement of the hand that he had let the bird fly away! On the morning in question the city of Lille had already been evacuated. Huntemann had stopped for a moment in the old Market Square and had told the prisoner that he was just "going round the corner to the post". The man understood and had made himself scarce by the time Huntemann returned "from the post". This was a highly individual method of solving a problem which had become acute and one which had my full support. We were rid of a useless man who had simply been forced to collaborate with us, a superfluous mouth to feed, and we reckoned that enough such men were already sitting behind barbed wire in Germany. . . .

On the afternoon of 1st September I drove over to Lüttich for a conference with No. III headquarters West to which I had been summoned. Unbroken double columns of vehicles were moving along all roads towards the east. The troops which had been sitting for years in France and Belgium were on the move. Order and discipline appeared non-existent. Each of these men was trying to place himself behind the supposed protection of the West Wall as soon as possible.

During the night of 1st September I received orders to withdraw FAK307 and all its subordinate formations from Belgium forthwith. It was to be regrouped afresh north of the Albert Canal and eastward of the Maas. After a terrible six-hour journey against the stream of the double columns moving in mad flight towards the east, I again entered Brussels at midday on 2nd September. I no longer needed any information about the true situation, for the roar of the engines of the retreating supply columns and the gun-fire of the English divisions advancing on Brussels from the south were too clearly audible.

On this Sunday, in spite of the lovely weather, the road from Tervieren onwards was empty of people. Single German vehicles were leaving Brussels at high speed and the last lorries were loading up in front of German headquarters buildings. Our own office in the Place de l'Industrie was still occupied by a small rearguard party which moved out after I left in the direction of Roermond. The second Staffel of the Kommando had already moved there in the early morning—it was expected that the enemy would occupy Brussels the same evening.

There was nothing more to be sought for in the city, but I was curious to see how the evacuation would be finally carried out. So with Willy and Ramm I made a short tour of several of the German headquarters which had ruled in this city for four years.

At the Luftwaffe Area headquarters we were met by heavy explosions—the destruction of the Communications Exchange, and a mob was looting the abandoned Hotel Plaza, the C.-in-C.'s headquarters. I noticed two men standing on guard with old 88 rifles in front of a private house and I recognised them both. They were two interpreters from Section IC of the staff of the Wehrmacht C.-in-C. Belgium and Northern France, who had been posted as sentries by their chief, Major von Wangenheim.

We had always had particularly good and friendly relations with Wangenheim. I ran in to say goodbye to him, and found him working on his papers as usual. When I asked him in surprise what he was doing, he replied that duty was duty and that he had received orders to defend Brussels "from all directions". To my further question about the forces at his disposal for carrying out this task, he drew me to the window and pointed with the gesture of a general to his two interpreters with their 88 rifles. There was not much to be said between us, as we had been all along in fundamental agreement about the events of the past six months.

As we started out at about 1700 on our journey to Antwerp a thick cloud of black smoke was standing above the lofty Brussels Palace of Justice. The mob had broken in for the purpose of burning the records of the police and of the Belgian Courts of Justice. Looting was in full swing, of buildings previously occupied by the Germans, soldiers' quarters, various messes, workshops and supply depots. Neither soldiers nor police were any longer to be seen.

The motor road to Antwerp was dead and deserted, and friend and foe alike seemed to have disappeared from the landscape. It was a lonely journey through no-man's land, which was not disturbed once, in spite of the lovely weather, by the fighter-bombers which ranged furiously along all the other Belgian roads. Hauptmann Riepe, the head of FAK Trupps, Ghent, had been warned and was awaiting me in Antwerp with a few of his officers. My impressions of the last few days were summed up in an admonition to them that order and discipline must be maintained by every possible means. They moved back during the night to Breda with orders to carry on their activity in the section between Antwerp and the Albert Canal.

I had given Lindemans his instructions for the future in Brussels before my departure for Lüttich. He was to remain behind in the city and establish contact with the Intelligence of the advancing English army immediately after the occupation of the capital. His active work as an enemy agent since 1940 and his connections with the Underground ought to speak well for him and should soon bring him into contact with the right quarters. We agreed that he should get himself employed by the English as a forward agent, so as to enable him to get back to us through the lines at the earliest moment possible. He was to report to the first German officer he met, who would pass him through to us in Driebergen via the IC of the nearest German headquarters. The IC of C.-in-C. Netherlands had been informed, and, on hearing the code word "Dr. Gerhardts" any individuals who passed through the lines would be sent to Driebergen. Whether CC would ever be able to carry out the task which I had given him was as dark and uncertain as everything else in these mad days.

We knew that Brussels and Antwerp were in a fever to greet the victors. So why should CC not now fight once again openly on the side of the enemy? What had defeated Germany still to offer him other than the certainty of being shot once his connection with me was discovered? What could I still offer him—I, a small Kommando leader in a hopelessly beaten army? I had arranged to make an advanced test of his future conduct by fixing up a last meeting with him for the evening of 2nd September at the Dutch Belgian frontier crossing-point at Wuestwezel. He might be able to get hold of important information for us in the interval.

We waited there for CC until midnight and, when he did not appear, drove on to Breda.

On the morning of 3rd September the Field Headquarters in Breda had a wholly peacetime appearance. It was impossible to obtain the slightest indication of what had taken place in Belgium during the previous night. Telephone communication with Antwerp had been interrupted since 0400, and no one knew whether the English had occupied Antwerp and perhaps crossed the Scheldt. At 1600 we succeeded in getting through to Hilversum, to find that the staff of the C.-in-C. Netherlands had no information about events on the Dutch–Belgian frontier. The IC in desperation begged me to attempt without delay a personal reconnaissance of the position in Antwerp.

Back, then, to Antwerp!

Only a few vehicles met us as far as Wuestwezel, and for the first time we crossed the frontier without being controlled—the guard posts were empty. All went well until we got past Merksem, when suddenly bullets began to whistle around our ears as we got to the middle of the large canal bridge in Antwerp harbour. Invisible marksmen, concealed in the large ware-houses on the south bank, had the bridge under fire. Groups of civilians and scattered German soldiers on the north bank were making their escape, springing to their feet one moment and seeking cover the next. We took shelter with the car in a corner among a group of houses on the south bank.

Fresh groups of men, women and children were continually hurrying past us, with occasional Germans—Luftwaffe and Marine—towards the bridge, where their March Hare antics commenced. I collected several NCOs and a Hauptmann with his right arm in a blood-stained sling, to give me a report. During the forenoon furious shooting affrays by the Under-ground had broken out at many points in the city. At the same time English armour had appeared from along the Boom road and had forced their way into the city, where German pockets were still holding out, after a brief engagement with a few flak guns. It was apparently not intended to fight a rearguard action, as no battle-worthy troops were available. The reports made up a picture which had become so familiar in France, and gave countless examples from the past few weeks of the same terrible description—standing fast for reasons of prestige, the frittering away of our strength, and losses of men and material without any hope of a decisive success. For weeks past no one had been willing to take the responsibility of giving the order for retreat. Hitler had retained the right to issue such orders for himself personally!

We dashed back across the bridge with the wounded Hauptmann in our car. The demolition detachment was standing ready to blow it up at the first appearance of the English. On the road north of Merksem there were marching several weak companies of Luftwaffe Field battalions without any heavy weapons. I discovered that they had orders to take up positions on the north bank of the Canal. Antwerp had been surrendered. . . .

In Breda the excitement of the populace could be felt. People stood in groups about the streets and leant out of the windows of their houses, staring southwards. German car after German car passed through. Surely the last German vehicle must soon have gone by and after it—the liberators! Were they really coming?

The gentlemen at Field Headquarters shook their heads over my report, which I asked them to pass immediately on to the staff of the C.-in-C. Hilversum. What? Is that a fact—Antwerp surrendered? No German troops on the Albert Canal? Nothing but a few companies of Luftwaffe? Brussels occupied yesterday and Antwerp today? In that case they may well be in Breda tomorrow! I shrugged my shoulders.

"If the Tommies don't stop of their own accord at the Albert Canal you can hand them over the keys of your safe tomorrow morning. All that stands in front of them on our side could not hold them for two hours if they were to cross over in force. Hauptmann Riepe's intelligence section is operating along the Canal. He has been ordered to maintain contact with you and report. And now, *meine Herren*, I must be on my way."

At dusk, to my relief, I found my IIIF party complete at Schloss Hillenrath. My anxiety on seeing the many burning vehicles on the road from Weert to Roermond and the screaming fighter-bombers, which ranged further and further in pursuit of the armoured columns, had proved groundless. I slept in the bed which had been occupied by the German Crown Prince on the first night after his flight from Belgium in November 1918. I slept deeply none the less—for I had not taken my boots off for four days and nights.

The confusion in Holland reached its height on the following day. The occupation of Antwerp by the enemy and the lack of battle-worthy troops on the Albert Canal had set countless German offices and headquarters in movement. All considerations of prudence seemed to have been swept away. The wildest rumours were extant and were believed. The offices in Nijmegen and Arnhem were to be packed up and loaded—all of them to be out by midday! Eindhoven was reported to be captured, and Hertogenbosch as well! One of the worst examples of the confusion of the day was provided by the airfield commander at Gilzer-Rijen. He abandoned it at a moment's notice with all his men, and left the airfield to its fate, the piles of ammunition by the flak guns and the well-filled stores of petrol and other supplies left unguarded. We posted sentries, and took the opportunity of replenishing our own nearly exhausted stock of petrol. Two days later a Hauptmann appeared from Hilversum with the object of taking over the airfield and stores in the name of the C.-in-C. The tragedy had its lighter side when he demanded back the petrol which we had taken!

But to return to "mad Tuesday". The signs that the Germans had completely lost their heads and the disgraceful disregard of duty shown on this 4th September 1944 caused a number of Dutchmen, carried away by their excitement, to take action which had unpleasant consequences later, when the English still didn't come and the Germans had after all to be reckoned with. In the meantime reports were received from our observers all along the Albert Canal that the enemy was making no serious attempt to cross it. Events hung in the balance for two or three days, and then it became clear that the enemy, too, was at the end of his resources. His victorious offensive from Normandy to the Maas had tailed off into exhaustion through a lack of organised supplies, lack of reserves of petrol and defective material. The first pause had arrived.

Quickly organised German formations, made up of reserve battalions and convalescent detachments, battalions of youths from the recruiting depots, with insufficient training, no experience and an entire lack of heavy weapons, were thrown into the line along the Albert Canal. We watched this spectacle with very mixed feelings. Were they still, in Berlin and at the headquarters of C.-in-C. West, not prepared to recognise that the defeat in the West was complete and decisive? Were they holding on to the chance of a recovery through a battle among the fortifications of the West Wall? Must the pointless final battle which lay just ahead of us really take place? We knew that no armistice was possible so far as Hitler was concerned, and we would have been ready to drag him down. But we also knew that we should go on fighting to the last cartridge if the enemy was going to insist on his demand for unconditional surrender. With a clear consciousness of being entangled in the web of a remorseless fate we went into battle for the last time, a battle which was to cost myriads of victims on both sides, which would lay flourishing provinces and cities in rubble and ashes, and whose issue in the long run could never be in doubt.

My FAK307 had been subordinated to Army Group B, which had charge of the defence of the section of the West Wall between Trier and Kleve, since the beginning of September. I withdrew the FAK troops under my command in Holland on 9th September to new duties in the West Wall. The Kommando headquarters was set up at Dersdorf, near Bonn.

A fresh chapter in my recollections begins with the new opponent who now faced us—the U.S. Army Secret Service—and with the altered interrelationship of the High Command in my own country. My task as head of the military counter-espionage in Holland, Belgium and Northern France had come to an end.

I can now round off the experiences of my period of duty in Holland by offering a contribution towards two matters which were closely connected with my activities and which have caused some heated discussion since the end of the war. These are "the betrayal of Arnhem" and the fate of the *Nordpol* agents. I shall confine myself to describing what I personally remember about these incidents from wartime, not feeling myself competent to discuss all the events which preceded them or to adopt any critical position.

FAK365, under the command of Major Wiesekötter, was separated from my command at the beginning of September, when he remained behind at the former headquarters in Driebergen to continue his counter-espionage work against the enemy advancing from the south into Holland. Wiesekötter was fully acquainted with the details of my connection with King Kong and of the task which I had entrusted to CC in Brussels. If this man were still loyal and was going to work for us he would have to come across to Driebergen. Wiesekötter would look after him, would let me know at once, and would if possible keep CC on the spot until I could arrive. The unexpected event took place late on the evening of 15th September. A report was telephoned to me in Dersdorf from Driebergen to the effect that CC had arrived, reported and left again with a new assignment—full stop! Nothing more! The hot-headed fellow had kept his promise, and I remembered the evening of my first conversation with CC in Brussels when he had said that he would show me what it meant to have King Kong for a friend!

Since Wiesekötter no longer came directly under me, there was no radio-link between us. It was in vain that I made every effort to get connected with Driebergen by telephone. I wanted to hear details, and as CC's personal chief I wanted above all to be able to influence his future activities.

On the night of 15–16th September all communications with Holland were interrupted. The great airborne landing of the English in the Nijmegen-Arnhem area had started—and it was to be a decisive failure. After weeks of heavy fighting the building of bridge-heads as jumping-off points for the planned Allied autumn offensive against the German North Sea coast was brought to nothing.

It was only at the end of the month that a written report reached me through No. III headquarters West from Major Wiesekötter about CC's visit to Driebergen on 15th September. King Kong must have understood perfectly how to gain the confidence of his new masters in Brussels, since on 14th September they had sent him through the lines to Eindhoven, with an order in his pocket for the head of the Underground there to stand by to go into action early on Sunday, the 16th. The main English attack was to start at this time. The English would meanwhile have crossed the Albert Canal on a broad front with the object of occupying Eindhoven and pressing on further to the north.

But King Kong had even more information than this! He knew that English and American parachute and airborne divisions were standing by in England for a big airborne operation which was just about to be started! I have never found out what King Kong reported to the IC of the German General headquarters in Vught when he was taken there on the afternoon of 14th September; but he did pass his knowledge on to Major Wiesekötter in Driebergen on the 15th. In any case there was no mention of Arnhem—King Kong had not mentioned it probably because he simply did not know in what area the airborne attack was going to be made.

The German Army headquarters in Holland, however, needed little imagination, quite apart from his report, to forecast the probable area of an airborne landing, though whether the report was evaluated in this sense is quite another question. To us CC was a concrete conception, but to the IC of the General Staff in Holland he was simply a suspicious foreigner who could well have been sent across for deception purposes. CC, in other words, was not the betrayer of Arnhem, simply because he was not in a position to betray it! But by passing his report to Driebergen he contributed towards the information which was obtained from other sources in the twelve hours preceding the English attack.

One of these sources was the Luftwaffe Radio Interception Service, which heard tuning-in at midday on 15th September by an RAF network which had not been previously encountered. "Tuning-in" can be described as the testing of aircraft radio sets in communication with a ground station before departure on an operation. This took place before every such occasion. The majority of the call-signs used by RAF units had already been identified by the German Radio Interception Service since 1942, and the list was being continually added to. Since that date the interception of tuning-in traffic had given excellent advance information of enemy air attacks before the squadrons concerned took off.

The traffic heard on 15th September was a conclusive indication of the imminence of a large-scale enemy air operation. Apart from this, on the afternoon of the 15th two English aircraft made a reconnaissance of some hours' duration over Nijmegen and Arnhem. This was not in itself particularly noteworthy, but what was significant was the fact that the two aircraft were given a very large fighter escort: during the reconnaissance, over thirty Marauders guarded the Nijmegen–Arnhem area from attacks by German aircraft. I heard that the German Command in Holland attributed this strong escort to the fact that a personal reconnaissance was being flown by the Commanders of the attacking divisions then being held in readiness in England. In the light of important observations such as these the reports of King Kong were of much greater confirmatory value.

On the afternoon of the 15th CC was sent at his own request by Major Wiesekötter to Eindhoven where he wanted to deliver the message to the local Underground from the English Secret Service in Brussels. We never saw him again, but we learnt shortly afterwards that his fate had overtaken him. An incoming cypher message was captured during the arrest of an agent-operator at The Hague. It was dated Brussels, 23rd November 1944, and explained why CC never again crossed the lines after the Arnhem affair. It read as follows *Achtung, achtung stop General warning stop Christiaan Lindemans of Rotterdam known as King Kong is a traitor working for the Germans stop He is under arrest stop Achtung, achtung stop General warning.*

The adventurous career of an audacious desperado had come to an end, a man who gambled with death for five years behind the scenes, and who recognised no law or obligation which he had not taken freely upon himself.

The whereabouts and the fate of the *Nordpol* agents gave me increasing anxiety after our retreat from Holland in September 1944, my attempts to retain some control over the fate of these men after the end of *Nordpol* having been brought to nothing by their transfer to Assen. The sharp deterioration of our relations with the head of the Sicherheitspolizei in Holland had caused us to lose all influence over their subsequent treatment. I did not doubt, indeed, that the RSHA would keep their

172

word and respect the lives of these men, but I was anxious to know the outcome for myself. For this reason I sent Huntemann from Dersdorf to The Hague to make enquiries. If he were to have a little luck with Schreieder he would be able to get some information on the pretext of wanting to speak with the agents about some operation or other.

Huntemann returned after a fortnight. He had been lobbying with Schreieder and also with different departments of the RSHA in Berlin, had been passed backwards and forwards, but had nowhere been able to get any reliable information about the whereabouts of the agents who had been moved from Holland many weeks previously to a German concentration camp. Finally he succeeded in having a conversation with Jordaan in the KZ at Oranienburg, whom he found ill and incapable of work. Lauwers, who was also in Oranienburg, had not been able to speak to Huntemann, and it had not been possible to confirm the whereabouts of the remainder.

We now knew enough. Our opponents would not let their prey escape again. The *Nordpol* agents had vanished into the grey mass of the millions of slave workers shackled to the oar-benches of the sinking ship of state, decimated by hunger and disease, awaiting the day of their liberation.

The knowledge lay heavily upon us of our temporary powerlessness against the robots of the system and against the forces driving us towards total disaster, which had become more alien, hostile and incomprehensible than enemies from another world. We had become foreign to one another in our thoughts and even unable to understand each other in our common mother-tongue. We had thought we were fighting for Germany, but had realised too late that, in our blindness, we had thrown even more than our homeland and our national existence into the scales of this war, and lost it: the heritage of the history of a thousand years, the cradle of human civilisation—Europe.

EPILOGUE BY H. M. G. LAUWERS

(During the war a lieutenant in the English Secret Service and operator of the clandestine radio-transmitter RLS—Ebenezer)

IT MAY cause surprise to find in an account by the former Chief of the German Military Counter-espionage of his war-time activities an epilogue from the hand of one of his defeated adversaries. I must therefore explain how this has come about.

In Holland after the war there appeared numerous publications dealing with the events described in this book. All of these were founded on German reports which became available to the authors. By reason of the lack of information from the "other side" they painted a picture of this drama as seen by the Germans alone. Although not in so many words, and however discreetly expressed, the impression could easily be given that the Dutch agents involved were not in fact up to their job. As the German *Nordpol* operation continued to function after each arrest, it was logical that the Germans should enjoy the full co-operation of the arrested agents. This should not, however, necessarily lead to the conclusion that these agents had not done their duty.

The German reports on which the above publications are based have, since the end of the war, led to Taconis and myself being branded, even in official quarters, as traitors who made possible the German success. Only my unexpected survival has enabled this legend to be scotched, at least in part. My fallen comrades cannot unfortunately speak for themselves. For their sake, I therefore consider it my duty to take advantage of the opportunity which has been given to me.

The agents who were sent out by MID-SOE between 1941 and 1944 were a very mixed batch. Some of them had reached England from Holland, others been brought to England from all parts of the world to help in the work of liberation. They came from all trades and levels of society, and represented every standard of education, from elementary to academic. All of them had one thing in common—a deep love of their country and, in their duties, a blind trust in their superiors. The long-standing reputation of the British Secret Service throughout the world and the training which the agents received brought their trust in their service to the heights of an almost mystical belief. Without this confidence not one of them could have been brought to undertake the dangerous tasks which lay before them. This boundless confidence considerably increased the effect of the shock which they sustained when they were dropped from aircraft straight into the arms of the enemy.

The method employed by the enemy on their arrival, of keeping the agents for hours under the impression that they were among friends, makes it understandable that they should have given away the agreed code-words which were intended to indicate to headquarters in London that they had arrived safely. This cannot give cause for criticism of the agents themselves. London asked by radio for these reports on the day after each drop into Holland. It therefore had to be assumed that an agent who thought he had arrived safely would try to pass the confirmatory message fairly soon after his arrival, without having to be asked for it.

Even if the Germans had not suggested it, a suspicion that there had been treachery in London must have entered the minds of these men. This suspicion may well have restrained some of them from using other means at their disposal for warning London, apart from the customary use of a code-word after their arrival. But from my knowledge of these men I am certain that many if not all of them must have tried to tell England about the disastrous actual state of affairs. I have by chance since learnt how opportunities were taken using either prearranged methods or special ones created by themselves. I am thinking of Andringa's warning to the agent Dessing and of a note which Akkie was able to smuggle out of Haaren. The escape of Ubbink and Dourlein was certainly not made only with the object of saving their own skins. Captivity at that time was less dangerous than escape, which would fairly certainly be unsuccessful. Their intention was to inform headquarters personally after all other means had apparently failed. The fact that England did not stop the game earlier cannot, as I have already said, be taken as a proof that the agents did not give any warnings. This statement may seem a bold one, but personal experience gives me a justification for making it. In order that the reader of this book may share this conviction, I will now relate my experiences in captivity. I have tried to fit in, as far as possible, with the events that Giskes describes in his book, but the attentive reader will note that my account differs at a number of points from Giskes' own. This is not surprising, as several of the occurrences described were to him simply unimportant details of his wider sphere of activity. But for me they were of such significance that they are printed indelibly on my memory.

The start of the *Nordpol* operation can be dated as Friday, 6th March 1942, the day of my arrest. As on every other Friday evening at 1830, I was trying to establish communication by radio with England. I remember well how cold it was in the Teller family's small parlour. I sat at my transmitter in a winter overcoat, with a blanket over my knees. The curtains were drawn, so that the glow from the valves in my set should not give me away. Close by the set, which I had just tuned in, lay three messages ready for despatch, and I now waited with my eye on the clock for the routine period to start. Nothing was going to come of it this evening, as before I could start transmitting Lieutenant Teller arrived with the news that a number of police cars were standing in the Oleanderlaan at the corner where it crosses the Fahrenheitstraat.

For weeks I had been reckoning on the possibility of being spotted by D/F. Only one of the places available to me was practicable from an operating point of view. This was the flat of the Teller family in the Fahrenheitstraat, and it was only here, where a long aerial had been rigged up for me, that I had succeeded in getting through to England. I had tried vainly to avoid such a source of danger, but was continually forced to return to this place, as the link with headquarters had to be kept open. Although the raid was unexpected, I was not, therefore, totally unprepared. While I quickly packed up the set, I racked my brains about what I could do. I did not at first assume that the situation was really serious. It was true that several police cars were standing outside, but a glance out of the window showed me that the street itself was not blocked. Clearly the enemy D/F detachment was not yet absolutely certain from where the transmissions were coming, as otherwise they would have broken in at once when they heard me tuning. I felt sure that we would be able to get away safely if Lieutenant Teller and myself were to walk quietly into the street and stroll away in close conversation. Frau Teller was given the task of throwing my case with the transmitter in it into the garden, in case the block should be searched.

I stuffed the three cyphered messages into my pocket and left the house without leaving behind any incriminating material except the set itself. Talking quietly, though our senses were at full stretch, we walked a few hundred yards down the Fahrenheitstraat towards the Laan van Meerdervoort without noticing anything unusual. The worst seemed to be over. We could not look round to see what was going on as we could easily have aroused suspicion by doing so. At the turning into the Cypressestraat, however, we were able to glance back. It seemed that our departure had passed unnoticed, for nothing unusual was to be seen.

We were walking somewhat more quickly down the Cypressestraat when we were suddenly overtaken by two cars. A dozen men shouting and waving pistols jumped out and surrounded us. Surprise was complete. In a moment we were forced back against the wall and searched for weapons—without result. Several weeks previously I had lent my revolver to Captain v.d. Berg and had not had it back. But what was now discovered were the three cyphered messages which I had brought with me in a feeling of false security, and which were now flourished triumphantly under my nose. I cursed my stupidity. Teller was pushed into one of the cars, and this was the last time I ever saw this splendid officer and good comrade. Sitting between two Germans in the back of the car, I was conscious that the possibility with which I had reckoned for five months had now become a fact. I was under arrest—a prisoner—and the game was up. The only chance of my getting back to England lay in escape, and the fact that I was now in enemy hands meant that my friends were in danger. Although I could myself no longer operate, one more important task remained, which was to be careful that my arrest endangered neither my comrades nor the work of the past months. My future conduct must be in accordance with the instructions which I had received during my training in England.

The objective of my involuntary ride proved to be the Tellers' flat, where on my arrival I found several Germans closely examining my transmitter, and others sniffing round the room in a search for suspicious material. I knew that their efforts would be in vain, as there was nothing more to be discovered. In any case, the situation was quite clear—my transmitter, the three cyphered messages and an obviously false pass were proof enough of my activity and my origins. I could not restrain a broad grin when an undersized police official showed a secret paper with great solemnity to his chief and threw me a triumphant look at the same time. As far as I could judge, it was a mathematical formula torn from a notebook belonging to Lieutenant Teller. The man looked angry and disappointed when a look of surprise and a scarcely discernible smile crossed his superior's face. Extraordinary, but this German clearly had a sense of humour!

This was my first meeting with Giskes. I was to learn to fear his acuteness during the following weeks and months, and to develop a feeling of distinct respect for him, albeit slowly and with reluctance.

My enemies did not make the first half-hour too difficult for me. There were no questions, simply a word of recognition for the courage of a soldier who had undertaken a dangerous job under difficult conditions. And although I had had bad luck, there was no need to worry about the future. It was not long, however, before this treatment changed, and attempts were made to profit by my shocked condition to clear up many details about my codes. I shall describe later and in greater detail how and why the enemy came to be successful.

When this first interrogation had had the result which the enemy desired—the giving up of my code—interest in my person became less. I was taken to the headquarters of the Sicherheitspolizei, where I was expected to provide further information. Everything now depended on whether I could avoid dragging Taconis or any other of my comrades into danger.

Soon after our arrival in Holland Taconis and I had discussed how we should behave if we were to fall into the hands of the enemy. We agreed that the one who was arrested must give the other at least twenty-four hours in which to go underground and secure the organisation. After that interval each would be free to act according to circumstances, but each must if possible keep silence about his comrades. Taconis and I had been dropped at the same time over the Stagerveld near Ommen, and our parachutes had been buried at the same spot. If after our arrest one of us was to admit to having been dropped from the air, and should both parachutes be found, which was highly probable, the enemy would become aware of the existence of a second parachuted agent. To overcome this difficulty we were going to say that we had arrived by torpedo-boat.

As the ostensible details of my arrival in Holland were already decided, I had now to fix the time of this arrival. In view of the desirability of deceiving the enemy about the purpose and objects of our Service in Holland, I must not name my director in England. It was plain that my statements would have to be acceptable, and not be contradicted by facts already in the possession of the SD. Luckily the enemy themselves helped me in this. During the interrogation at the Fahrenheitstraat one of the Germans, who was, as later appeared, Leutnant Heinrichs, related that they had intercepted my first transmission to England on 3rd January 1942. He remarked to one of his assistants that I might well be "one of Derksema's people". When in England I had by chance had some contact with an intelligence service to which Captain Derksema belonged. In conversation with an agent called Tazelaar in Holland Taconis and I had heard that Captain Derksema had left the Service and that a new organisation had been created. I hoped now to make use of this knowledge in appropriate fashion.

During the days and nights of arrest which followed I dictated statements for hours at a time to Kriminal-Sekretär Bayer. I can set down in a few lines the main points of these fairy stories. According to them I had been landed on New Year's night 1941 from a torpedo-boat on the coast at Nordwijk in the pay of the Dutch Intelligence Service. In accordance with instructions given to me in England I had had a conversation at an Amsterdam café with an unknown individual whom I never saw again. This man's business was to hand me the messages which I was required to transmit. By this I succeeded in my purpose of not mentioning by name either my director in England or my contact in Holland. My report was full of interesting details and seemed to have made a good impression. Bayer, however, came back three weeks later and complained to me that it "was all lies".

I must now say a few words about the training which was given to embryo agents in England, so as to make later developments easier to understand.

There were a number of different training schools. Every potential agent had to visit a series of these establishments in order to be fully conversant with all the aspects of his trade, after which he was trained for his special task. As a result of the shortness of the time available it was not possible to visit all the schools, nor, on account of the specialisation, was it really necessary to do so. But one school had to be visited by everybody. This was the Security School, where methods were taught by which one could protect one's person, code and organisation, etc. In contrast to other schools, where most of the instructors were British Army officers, the veil covering the work of the Secret Service was here lifted by men who had already themselves acted as agents for longer or shorter periods. Quite naturally the possibility was discussed of an agent falling into enemy hands, what the enemy would be interested to find out in such circumstances, what means he would employ to obtain this information and how the agent should conduct himself.

The head instructor explained the enemy tactics as follows. As his first object, the enemy will try to get hold of the code, beginning with friendly talk, good treatment and promises, then with threats and finally through efficient maltreatment. But while the instructor recognised the fact that we might be determined not to reveal the code, he indicated that the Gestapo had means at its disposal which can break down the resistance of the strongest man. By the end of it the enemy would not only have learnt our code, with all the consequences that would ensue, but our weakened resistance would then cause us to reveal things of greater importance such as personal contacts and the security check. Our only means of informing England of the fact that we were under arrest was this security check, and it was for this reason that it must in no circumstances become known. The conclusion was that we must avoid a third-degree interrogation if at all possible and should surrender the code reluctantly by bowing to superior force.

The enemy would then use his knowledge of our code to play the transmitter back and would try to obtain our co-operation in order to increase his chances of success. In other words, an attempt would be made to "turn us round" and to use the set in the enemy's interest.

The head instructor gave us to understand that although we would certainly feel in honour bound not to help the enemy in this way, such conduct would be as unwise as it was praise-worthy. The enemy would have no chance of success in playing-back a transmitter so long as he was not in possession of the security check. He might try to play the set back without our assistance. If we pretended to help, we could acquire considerable credit in the enemy's eyes and could gain time for a possible attempt at escape.

Such a development would not be of much advantage. The Allied High Command was interested above all in establishing a radio-link with the enemy which would be considered as being above suspicion, but such a link could only escape suspicion if the enemy were under the false impression that the operator had been completely "reversed".

After my arrest, these statements by my instructor turned out to be correct. The attack on my code began straight away in the Fahrenheitstraat. Heinrichs referred to many particulars of the code system which I used and declared that he could, without my help, decypher the reports which he had found on me. He wanted to give me an opportunity of saving my skin by handing over the particulars of my code voluntarily, and he added that I could save him a lot of trouble by doing this. To me it seemed reasonable to meet this proposal, and I promised that I would fall in with his wish if he were to succeed in decyphering one of the three messages which had been found on me. To my surprise he agreed at once. He sat down at a table, seemingly immersed in his "game of patience", and after about twenty minutes declared triumphantly: "I see

—*the cruiser Prinz Eugen is lying at Schiedam*—eh?" This was in
actual fact the text of one of the messages!

I freely acknowledge that I was taken completely by surprise
by this unexpected development. I only found out the full story
when I discovered later that the report had come from Giskes
himself, who had passed the information to me through Rid-
derhof and v. d. Berg simply in order to make certain that I
would have something to transmit on that Friday evening.
Now the text of the report was being used to break my code,
and I had to keep my word that I would hand over the code
system itself. The enemy had, however, no idea that I was
doing this with the approval of my superiors.

My self-confidence was not shaken during the first few days
after my arrest. It was in fact so great that it gave me some
trouble to play the part of a man taken completely by surprise.
My confidence in Taconis's help contributed towards this,
provided he succeeded in evading the enemy. Herr Nakken,
my landlord at The Hague, had not been arrested. In accordance with our arrangements he would have warned Thijs
(Taconis). As the SIPO appeared to believe my story, I had
no immediate anxiety about his safety.

But my self-confidence suffered a severe shock when, three
days after my arrest, I saw my two friends Thijs-Taconis and
Jaap van Dijk pass the door of the room where I was being
interrogated, in handcuffs, particularly as I could find no
explanation for their arrest.

This fresh shock was a severe one, and my dismay was even
greater when Giskes said suddenly as he left: "And what kind
of mistake do you have to make?" This was when I had already deciphered the three signals which had been found on
me and after I had asked for twenty-four hours in which to
think it over and make a decision about his proposal for further
collaboration.

I knew that the enemy was fully informed about the code
system in use by us. After the training which I had received in
England this did not surprise me, and I was not therefore
astonished when Giskes let me see that he knew that I must
be in possession of some kind of security check. A routine
question such as this was to be expected from a specialist in
counter-espionage. But my heart had stopped for a second

when he appeared to have knowledge of the actual kind of security check which was in use.

It now seems the right moment to go into this question of my security check more closely. This check was a means of indicating to my director in London that all was well with me. A lack of this check or any unexpected change in its nature would indicate that the agent concerned had fallen into the hands of the enemy. Any arrangement made for the purpose— even the simplest—will fulfil its object if it is rigidly laid down that normal use proves that the agent is working in freedom, and absence or alteration of the check in use that he has fallen under enemy control. The check-indicator, furthermore, must be selected in such a way that it cannot be detected by an enemy operator listening in to the transmissions. It is quite obvious that headquarters must keep strictly to the prearranged plan, as otherwise it is better left altogether. The check, when properly used, will provide headquarters with complete and unbroken control, and its absence or alteration should result in a breaking-off of communication with the agent concerned, or at the very least in regarding the link with suspicion and in setting in train a searching investigation. The agent's duty, therefore, is to use his security check with the greatest care and to leave it out of his signals directly he is arrested. If the check is carefully chosen the enemy may think that he knows it, but will in fact never discover it. If pressure should be brought to bear on the agent, he must not reveal the correct check. The enemy will be bound to accept any statement which he may make up on the spur of the moment and will thus not be given the possibility of establishing control.

My own security check, as arranged in London, was related to my agent number—1672. The first two figures—16— indicated that I was to make a deliberate mistake in every sixteenth letter in the text, in such a way that this mistake would not be affected by the accidental addition or omission of a single dot or dash during transmission. For example, an S ($\cdot$ $\cdot$ $\cdot$) must not be altered to an I ($\cdot$ $\cdot$) or to an H ($\cdot$ $\cdot$ $\cdot$ $\cdot$) but into a T($-$).

As his question had indicated, Giskes was aware of the fact that our usual security checks consisted of deliberate mistakes made according to a prearranged plan. This would not have caused me so much disquiet had not the three decyphered messages, which included the security check, been discovered when I was arrested. With the knowledge available to Giskes it would now be possible for him to work out my check from a study of these messages.

It seemed that the game was up!

My first reaction to Giskes' remark was to pretend that I did not understand what he was talking about. I could gain time in this way, but each subsequent routine period would serve to bring the true facts to light. The twenty-four hours' grace which I had asked for now came in useful, for I came by chance on a solution of the problem which might well save the situation. In two of the three messages found by the enemy my security check had coincided with the word "stop", which occurred frequently in the text. If, when cyphering up the message, I had inserted the check in every sixteenth letter, as I usually did, it would have mutilated the text too greatly and made the meaning too obscure. It had, quite by chance, been possible for me to introduce my check into the word "stop" in both messages. So I changed one "o" into an "i" and the other into an "e", which avoided the mutilation difficulty. I decided, therefore, to tell the Germans that my check consisted of changing the word "stop" once in every message into "step" or "stip". If this were to succeed, England would know, not only that I had been arrested, but also that the enemy had become aware of our check system. The check which I had invented would, of course, not stand too close a scrutiny, but, if it were discovered that another mistake had been made in the third message, I hoped to attribute this to an oversight. And to stop the false check from being accidentally changed into the correct one in the messages which I was passing for the enemy, I had to try to get the actual cyphering into my own hands.

To cut my story short, my "check deception" was successful, and all signals which I transmitted over the transmitter RLS-UBL from 12th March to the end of October 1942 were cyphered up by me personally, so that I can confirm that each one of these messages contained a clear warning.

During my second routine period in the service of the enemy, the first in which the correct security check was absent from my signals, the German Abwehr received information that a new agent called "Abor" was to be dropped in shortly. A number of containers of weapons would be dropped at the same time, also a piece of apparatus which was in no circumstances to fall into enemy hands.

It was little short of disastrous that the enemy should receive this information before the fact of my capture had become known in England, as the first of my messages which did not contain the security check had not yet been decyphered over there. In these circumstances it was practically impossible for London to be able to deceive the enemy with success, and I was fully prepared to see the radio-link cut short abruptly. The confidence which I thought I had gained would immediately turn to the opposite extreme, and if Giskes were to have my security check investigated again I could have no illusions about my subsequent fate.

As a result of these considerations, which were passing through my head while I was decyphering the message, I went over at once to the attack and refused any further collaboration. To let a few containers of weapons fall into enemy hands was one thing. But to betray comrades who were coming over from England? No—never!

This sort of behaviour may have seemed self-evident to the enemy, but it was in fact pure make-believe. I had no need to worry that Abor might fall into the hands of the Germans, since, before he could be sent across, it must certainly have become known that the Germans were sitting behind my transmitter in Holland.

My immediate object was to prevent the unfavourable impression which would be made should the radio-link be broken off by London. An agent would seem to be deserving of Giskes' confidence if he appeared to be convinced that his collaboration would bring his comrades to disaster and if he firmly refused this co-operation as a result without considering the possible consequences. Unfortunately, this *ruse de guerre* was never actually put to the test.

Back again in my cell, alone with my thoughts and with all the time in the world for melancholy fears to enter my head, doubts began to assail me. Would England indeed react to the omission of my security check? Before my departure I had even reminded my accompanying officer on the way to the airfield that the smallest deviation from my security check would show that I had been captured! And if this deviation should have been overlooked, what then? Still, this was clearly nonsense, as the check was of much too great importance! It was a relief when Giskes appeared once more to tell me that my co-operation was necessary at that very moment to enable him to save the life of the agent about whom we had been forewarned. I did not give my consent at once, but nevertheless at the next routine period I found myself sitting again behind my transmitter, intent on getting an answer from England to my calls. And England answered. I passed two messages to London and received one in return.

I started anxiously to decypher it and felt myself gradually relaxing as the first readable lines of the message came out. It appeared that the agent was ill—but that the drop of material would take place as arranged. From my point of view this could only mean that my warning had been received and understood. *England* was now going to start deception on her own account! But then all my doubts returned. Abor might indeed be ill, but what about this important piece of apparatus of which we had been informed? To restore my confidence I managed to persuade Giskes to let me have a look at the captured material under the pretext that I might be able to provide some information about the contents of the container. The material which I was shown consisted of a small quantity of weapons, some unimportant pieces of sabotage material, and a few packets of leaflets printed in French. My own satisfaction was considerably more than the Germans'!

Soon afterwards the blow fell. After a considerable number of messages had been passed to London without the use of the security check Abor was dropped on 28th March 1942, together with the piece of apparatus about which we had been told. When London signalled that Abor had left, a feeling of stupefying bewilderment came over me. I had been too sure that headquarters would react correctly to the absence of my security check.

At that time the situation in Holland was such that only a "blind" drop, that is to say a drop without the use of a reception party, could ensure the security of both the agent and the organisation concerned. Even if they had been amateurs it could hardly have been expected that they would send across an agent in all good faith using the medium of a suspect radio-link, and in my mind I could not reconcile such bungling with my knowledge of the British Secret Service. Just one conclusion remained—that if the agent really did arrive it would be in accordance with some special plan made by headquarters.

But supposing I was mistaken? Supposing I had had a mental lapse or had overlooked something? I simply could not believe that I had been misunderstood in London. And I could not take it on myself to remain passive in face of a development such as this. I could only achieve complete security if England stopped the "game". And I could only assume that everything was in order if I were to succeed in getting yet another warning across to headquarters, apart from the missing security check—even if the "game" went on. Having reached this conclusion it simply remained for me to consider how I could bring effect to this plan. A rapid achievement of my object could only be attained through my set. I must try to transmit either a single word of warning or some indication in the text of a message. In order to make such a warning comprehensible several symbols would be necessary and I could only assume that the two German operators who supervised my work would take immediate action if I should depart from the normal procedure signals or from the correct cyphered version of the message. Should this occur I should neither have obtained my object nor have another opportunity of repeating the warning. Even if I were to succeed in giving an unmistakable warning before I was stopped the enemy would know that England had received it, and in that case all London's chances of an undetected deception operation would have finally gone. Although I felt it right to transmit another special warning, my decision must not interfere with the plans of headquarters. I must handle it in a different way—unnoticed by the enemy and yet clear enough for the English operator to understand.

Clandestine radios operated at that time in exactly the same way as commercial radios. In this way the English clandestine radio authorities hoped to prevent the traffic from coming to the attention of the enemy. In the so-called commercial Q-code which was in use each operating signal is of three letters beginning with the letter Q. Special operating signals were used for changes of frequency only. The agent operator worked in all other respects in the same way as his colleague, the commercial operator, and the headquarters station had therefore to watch very closely the working of each line and to report the slightest unusual feature or deviation from the normal. In order to prevent the operator at headquarters from overlooking any deviation, it was strictly laid down that the entire traffic must be taken down letter by letter, as I well knew. Every agent transmission was read by two operators in England independently of one another as an additional safeguard. Even if the absence of my security checks had given headquarters no cause for suspicion that I was under arrest, I could at all events count on great watchfulness, and the following plan was based on this assumption.

The Q-code signal QRU is used during every radio transmission to indicate "I have nothing further for you". I resolved in future to transmit CAU in place of this signal. The change in the rhythm of transmision of the first letter is very noticeable, in the second letter it is one dot only, and in the third letter it is unchanged:

QRU — — · —   · — ·   · · —
CAU — · — ·   · —   · · —

I hoped to be able to transmit this changed version without drawing the attention of my watchers. But the English operator, who would be expecting QRU, could perhaps also miss this small variation. However, in order that the man at the receiving end might check what he had written down the first time, I proposed to make the customary repetition at each group of letters slowly and with emphasis, so that a normal degree of watchfulness in England must soon make the variation plain.

Having transmitted CAU in this way it would now be necessary to add the letters GHT, and this could only be done through the medium of the frequency-changing signals, so as to spell out the word CAUGHT. My particular signal LMS was not known to the German operators, as it was not an operating signal in normal use. This alteration, however, would certainly call attention to itself in England. If both the alterations had been noticed even once no black art would be necessary to construct the word CAUGHT out of the two groups.

Abor had already been disposed of before I could bring this plan into operation. Headquarters had informed him about my radio-link, assuming as they did that he would be in safe hands, but he had nevertheless to be supplied with a foolproof method of warning. Headquarters never made any exception from this rule, as I had been taught in the Security School. This also gave me hope that the deception being practised by the wrong people would now be brought to an end.

But the radio traffic continued normally even after the arrival of Abor. I made a habit of changing my transmission frequency when the circumstances did not require it and even when it was not desirable, always using an operating signal which I would alter to GHT. I sent my CAU at the same time regularly into the ether without the German operators noticing anything. Their watchfulness indeed sometimes increased noticeably after I had transmitted CAU, but never as a result of the altered frequency-changing signals. I had evidently won their confidence to such an extent that they harboured no more serious suspicions.

At this time I had no anxiety as regards the final outcome of the "game". I learnt in conversation with Huntemann that a number of new agents must be operating in Holland. It seemed clear that headquarters was still keeping to the agreed principles of only dropping agents in "blind". There must, it seemed, have been special reasons behind the exception made in the case of Abor.

I also learnt of the presence of fresh agents in Holland through the messages which passed through our hands. One day headquarters signalled that agent "Pijl" would be calling at the address which was known to us. The object of this signal was a riddle, as it must now have been known for some time that my set was in enemy hands. However, Pijl's permanent address, which before my arrest we had confided to the agent Homburg who had since returned to England, was not known to the enemy, so that he was not running into any particular danger. I believed, too, that Pijl himself would warn England when he found that he was unable to make contact with us. At the next routine period headquarters signalled that Pijl had been unable to make contact and asked us to say where he could find us. Giskes had no need to take action on this, however, as Pijl had already fallen into the enemy net!

After Pijl's arrest the catastrophe was complete. All the MID-SOE agents who were still at large at that time were arrested in the course of a few days, except for one called Dessing.

I will not try to describe how I was torn between my beliefs and my doubts in the months which followed, as I have not the ability to clothe these feelings with words. I will confine myself to the statement that I was able to send further warnings to headquarters. In trying to explain clearly how I did this I must first of all explain the method of cyphering up signals.

The principle was a very simple one. The letters of the message were systematically interchanged, though themselves remaining unaltered, and the recipient had to work backwards, using the same system, to get the plain language text. I will use the following message as an example:

Nr. 36 waarschuwden alle betrokkenen stop gevaar voor organisatie klein stop einde.[1]

This text is written horizontally in the form of a square, the letters falling directly one below the other. The resulting vertical columns are then numbered in accordance with an agreed number, the so-called "key number". In order to make decyphering more difficult for unauthorised persons, five to ten additional letters are added at the beginning and end of the text. Using this process, the above message will read as follows:

5	I	10	4	9	8	2	7	11	6	3	12
C	G	H	V	H	G	C	R	N	R	D	R
I	E	Z	E	S	W	A	A	R	S	C	H
U	W	D	E	N	A	L	L	E	B	E	T
R	O	K	K	E	N	E	N	S	T	O	P
G	E	V	A	A	R	V	O	O	R	O	R
G	A	N	I	S	A	T	I	E	K	L	E
I	N	S	T	O	P	E	I	N	D	E	H
G	C	H	C	G							

[1] *Have warned all concerned stop Danger for the organisation small.*

Columns are then read off from top to bottom in the order in which they are numbered, and written down in the form of another similar square as follows:

5	1	10	4	9	8	2	7	11	6	3	12
G	E	W	O	E	A	N	C	C	A	L	E
V	T	E	D	C	E	O	O	L	E	V	E
E	K	A	I	T	C	C	I	U	R	G	G
I	G	R	S	B	T	R	K	D	R	A	L
N	O	I	I	G	W	A	N	R	A	P	H
S	N	E	A	S	O	G	H	Z	D	K	V
N	S	H	N	R	E	S	O	E	N	R	H
T	P	R	E	H							

All the letters are now systematically displaced. The signal is already unreadable and it can be encyphered afresh in a similar fashion should additional security be required. Finally, the columns are again read off vertically, and the letters written down once more horizontally in groups of five. Another key number is then required to enable the recipient to discover the order of columns which is being used. Exact knowledge of this key number is essential for the rapid decyphering of such messages and for this reason the number is repeated at the end of the message. Quite apart from this, the cyphered version is given another current number, for use with the recipient, also two figures giving the number of five-letter groups in the message and the total number of letters contained in it. The final form of the encyphered message will then look like this:

Nr. 36 *gr.* 18 34512 *etkgo nspno crags lvgap krodi*
siane gvein sntae rradn coinkn hoaec twoee ctbgs
rhwea riehr cludr zeeeg lhvh 89 34512

I have described this method of cyphering and the final form of the signal in some detail, as they played a large part in my subsequent warnings to London.

I wanted to make one more attempt to transmit the word CAUGHT, but in such a fashion that the operator who received it in England would be bound to write it down as one word. I hoped to do this by changing a suitable group of letters in a message. In order to increase the chances of having a suitable group at my disposal I used the consonants C, G and H almost exclusively for supplementary letters. And I very soon achieved the desired results. I have included in the above complete cypher message an example of how this warning was passed. In this example the group CRAGS became the third group by the use of C, G and H as additional consonants. This group now had to be changed to CAUGH, and a comparison of the two groups in Morse code shows how easily this can be done by the addition or omission of single dots:

c	r	a	g	s
— · — ·	· — ·	· —	— — ·	· · ·
— · — ·	· —	· · —	— — ·	· · · ·
c	a	u	g	h

T had to be added to the group CAUGH and I transmitted this as the first letter of the following group. As I have already mentioned, each group of letters was made twice, except under very good reception conditions. The foregoing example was therefore transmitted as follows: NR 36 NR 36 GR 18 GR 18 34512 34512 ETKGO ETKGO NSPNO NSPNO CAUGH CAUGH T(short pause, to give the receiving operator a chance to write down all that he had previously heard), erase sign, GR 3 GR3 CAUGH CAUGH T (another pause and an apology to the German operator who is now listening attentively), erase sign, CAUGH CAUGH T (a final pause and several irritable longs on the key, at which I curse my clumsiness with the object of deceiving the German operator, who is growing uneasy), erase sign, GR 3 GR 3 CRAGS CRAGS LVGAP LVGAP . . . . . . LHVH LHVH 89 89 34512 34512.

In the above-quoted example the warning may not seem particularly clear, but the picture becomes very different if one imagines what is written down in the log book of the receiving operator in England. As the erase signs are not themselves written down, the message would have appeared as follows:

NR 36 GR 18 34512 ETKGO NSPNO CAUGH T GR 3 CAUGH T GR 3 CAUGH T GR 3 CRAGS LVGAP . . . . . . LNVH 89 34512

and a stroke must be imagined through each group CAUGH T.

EPILOGUE

The threefold transmission of the same mistake in one group and the fact that the following group did not really begin with a T, which amounted to the transmission three times of an alarm signal CAUGHT must have made it clear, at least in my view, that there could be no question of a normal radio transmission. At the end of the message London asked for a repetition of the third group and my reply was acknowledged with the signal "understood"! This was a final reason why I was convinced that my warning had got through. I came to the same conclusion by reason of other considerations. I had transmitted the warning CAUGHT three times quite shortly after London had given orders that preparations should be made for the demolition of the Kootwijk transmitter. We were instructed several times to wait for a special signal to be transmitted over the BBC before starting the operation. I therefore remained in hope that a separate and larger British operation might have been called off as a result of my warning, until one day the order to blow up Kootwijk was actually received.

At the end of July 1942 I was moved into a cell in Scheveningen which I shared with the agent Jordaan, who had been arrested at the beginning of May, and I learnt from him that he had given away his security check to the Germans. He reproached himself bitterly with this and I listened with close attention as he described the circumstances which had brought it about.

Jordaan had had the duty of operating for the agent Ras, who was the organiser of his group and who had disappeared suddenly, without trace, in Utrecht at the end of April 1942. Jordaan had not been able to find out definitely whether Ras had been arrested, and had proposed to headquarters that he should hand over his set to an operator who would be recruited on the spot, so as to leave him free to continue with the building up of the organisation. Before the new operator could do a test transmission, however, Jordaan was himself arrested. The key of his code was found on him, and a German operator took his place without either knowing or using his security check. Headquarters approved the new operator as being efficient, and at the next routine period told Jordaan to "instruct him in the use of his security check". . . . It was in

194

this way that Jordaan, rendered helpless by this order, had come to hand over his check.

We had to try and make good this mistake, which had been so understandable in the circumstances, and we saw a possibility of doing this when one day London requested a written report which was to be sent across by a courier route via Sweden. I was to attach a letter in my handwriting for authentication purposes, and the Germans gave me instructions to write this letter. Before my departure from England I had been given an address in Switzerland through which I could communicate back should for any reason my radio-link break down. If everything was normal I was to write an ordinary chatty letter and sign it "Wim". If, however, I were to sign such a letter "George" it would indicate that I was in serious difficulties. I wanted to make use of this in my letter to Sweden, but this did not come within the original arrangements.

In the Security School we had been taught how to introduce short disguised messages into the text of any letter, which could not be detected by the uninitiated. Up to that time I had not written a letter of this kind to headquarters, as no instructions had been given me to do so. Jordaan, on the other hand, had been given such instructions before his departure, so that we might be able to smuggle a message back in his letter code. But how were we going to get head-quarters to look out for a message cyphered in Jordaan's code when it would be contained in the text of a letter written by me? But we thought we could solve this problem, too. The wording of a message simply provided the framework for the coded portion, and the text itself was no guide to it. If in this instance we could write the letter in such a way that it could only originate from Jordaan we thought that we could gain our object. I put at the end "Kindest regards to Miss X" and asked that she should be told that I still had the pleasantest memories of the hours which we had spent together in the restaurant "Y" in London. The lady concerned was quite un-known to me, as she had joined our Service months after my departure. But Jordaan knew her well and had often taken her out to this particular restaurant. As a consequence of reorganisation at headquarters I knew scarcely anyone on the staff at that time, and no one knew me personally. My signature "George" would indicate that something was wrong, and the regards to Miss X ought to lead the scent on to Jordaan. It should also lead to the conclusion that Jordaan and I were in contact with one another, and this would contradict the situation reports about Holland which had been sent across by the Germans. We thought that we could count on London to take warning from the mines which we had planted in my letter.

But our efforts were in vain. We heard from Huntemann months afterwards that after a first attempt to pick up correspondence from a "post box" in Delfzijl, which had gone wrong, headquarters had made no further attempt to get hold of my letter.

Meanwhile we tried our luck again over the radio. We decided not to confine ourselves any longer to the transmission of an alarm word, but worked out a method by which we hoped to pass the following text with the help of the supplementary letters used in the cyphering process: *Worked by Jerry since March six (th) Jeffers May third*. (Jeffers was Jordaan's cover-name.) It was clearly impossible to include this text in the supplementary letters of one single message and it had therefore to be split up. Furthermore, I had been forbidden by the Germans to use vowels among these letters. Being restricted to consonants meant that individual letters would have to be changed during transmission in order to make the warning text come out clearly when decyphered in London. One difficulty was in finding the letters again which were to be changed, in the cyphered form of the message, but I solved this by writing down these letters in a particular way. In the following example the letters to be changed are shown in bold type:

8	4	5	1	7	13	10	11	2	9	12	6	3	14
W	**G**	R	K	**T**	D	B	**K**	N	R	D	R	I	E
Z	E	S	W	A	A	R	S	C	H	U	W	D	E
N	A	L	L	E	B	E	T	R	O	K	K	E	N
E	N	S	T	O	P	G	E	V	A	A	R	V	O
O	R	O	R	G	A	N	I	S	A	T	I	E	K
L	E	I	N	S	T	O	P	E	I	N	D	J	T
R	R	**K**	S	S	N	C	**T**						

8	4	5	1	7	13	10	11	2	9	12	6	3	14
K	W	L	T	R	N	S	N	C	R	V	S	E	I
D	E	V	E	J	G	E	A	N	R	E	R	R	S
L	S	O	I	K	R	W	K	R	I	D	T	A	E
O	G	S	S	W	Z	N	E	O	L	R	R	H	O
A	A	I	B	R	E	G	N	O	C	K	S	T	E
I	P	T	D	U	K	A	T	N	D	A	B	P	A
T	N	E	E	N	O	K	T						

NR57 GR19 31414 TEISB DECNR OONER AHTPW
ESGAB NLVOS ITESR TRSBR JKWRU NKDLO AITRR
ILCDS EWNGA KNAKE NTTVE DRKAN GRZEK
OISEO EA 92 31414.

In this message seven letters had to be transmitted in altered form, a G as an O, two K's as Y's, an S as an I and three T's as E's. This would not be difficult to remember and should be quite feasible. After decyphering, the message would come out as follows, the first part containing our warning:

8	4	5	1	7	13	10	11	2	9	12	6	3	14
W	O	R	K	E	D	B	Y	N	R	D	R	I	E
Z	E	S	W	A	A	R	S	C	H	U	W	D	E
N	A	L	L	E	B	E	T	R	O	K	K	E	N
E	N	S	T	O	P	G	E	V	A	A	R	V	O
O	R	O	R	G	A	N	I	S	A	T	I	E	K
L	E	I	N	S	T	O	P	E	I	N	D	J	E
R	R	Y	S	I	N	C	E						

But now another difficulty arose. The two operators who had controlled my transmissions for months past, and who seemed to have confidence in me, had been suddenly succeeded by a testy official, an older man, who watched my every movement with the eye of a hawk. When cyphering he read out the letters to me so that I could not make a note of those which I had to change during transmission. I had practised our system over and over again in the cell with Jordaan, and when one day he was able to come with me to the transmitting-room I pushed the paper over to him and asked him to help me. For a moment it looked as though the German would interfere, but Jordaan had already started to read out the letters to me, and he let him continue. Jordaan was then able to indicate which letters were to be changed by short pauses and a cough at appropriate moments. The first part of our warning was passed in this way.

I do not know whether the controlling official had noticed something, or whether the decision was taken for other reasons, but from that day Jordaan was never again allowed to accompany me to the radio-room, and there was therefore no longer any possibility of passing the rest of the warning message. But we still believed that we had succeeded in our object when headquarters told my station RLS–UBL to make contact with other stations in Holland. In our opinion this move could only have been made for purposes of a check.

For weeks at a time only unimportant signals were interchanged over my set. Drops which had been arranged were cancelled or postponed. The same applied to the other transmitters, and our confidence increased as the faces of the Germans grew longer. But we once again suffered a terrible disappointment when one day I received a message which announced the departure of two new agents.

Jordaan and I, in our despair, tried to work out some explanation for these happenings. We had realised that a catastrophe was possible, and our warnings had been sent as a result, but in the depths of our hearts we had retained complete confidence in headquarters. The mistakes which were being made by London were so utterly at variance with the basic principles of security as we had been taught them that it seemed incredible that they could be genuine. We could believe, at a pinch, that my attempts at sending the word CAUGHT had not been successful, my letter had perhaps not yet reached England or had simply been pigeon-holed, but my latest warning had been too plain for anyone to have been able to overlook it. Was it not more explicable to accept the solution that headquarters was carrying out a widely spread deception operation? Had not the English been masters of this technique during the First World War? One day an invasion of the Continent would be mounted from England. Perhaps a number of agents were now being sacrificed in order to establish contacts with the enemy which would support such an undertaking and hold out a prospect of success. We simply did not know. There was nothing for it but to search for new methods which would preclude any possibility of misunderstanding.

This could only be managed in one single way. A message from us would have to be passed in the place of a German one. Control was stricter than ever, and before we could complete our preparations Giskes had already decided to have me replaced by a German operator.

One of the last messages which I sent for the Germans was a proposal that a relief operator should be brought in instead of me, ostensibly in order that I could be free to help Taconis in his work. London had already approved the introduction of relief operators on other lines and gave instructions for the new recruit to carry out a transmission test. My own controller did this test—an old hand who had probably something like twenty-five years' service behind him. But there was a surprise when London gave instructions for the link to be operated as previously, "as the new operator was not suitable!"

The Germans paid not the slightest attention to these orders. The man who had been turned down continued operating without any further protest from London, and in this way my part in the *Nordpol* operation came to an end.

London's decision that I was to continue operating astonished us all. Why was no relief operator now acceptable? Up to the time of my arrest, although I was an amateur I had perhaps done my work as an operator well. Later on, by incorrect coding and mistakes in the cyphered text, I had made some intentional blunders. Even at my best I could never compare with the German professional operator who was to succeed me. Why then this incomprehensible decision from London?

Once again I thought that I knew the truth.

England was aware of my position and was trying to have me kept on as an operator in order to ensure my safety. At the time when Jordaan and I were transferred from Scheveningen to the Haaren camp in Brabant in November 1942, we had been convinced that the game which was being planned and played by the German Abwehr had been seen through in England and was in fact being controlled from over there.

Today, years after the end of the war, when the whole extent of the catastrophe has been revealed, an idea like this must seem a little naïve. My only explanation is that it was born of unshakable confidence—a boundless faith in the Service to which we belonged.

It is difficult to describe the atmosphere which had spread its all-pervading influence about us during the period of our training in England. Jordaan and I had seen how the struggle was being carried on there when the enemy had already conquered all Europe. We marvelled at the determination with which the English people had withstood terrible blows and yet made preparations for a future counter-stroke with unbelievable tenacity.

In this great country there existed Secret Services which were fighting before the first shot had been fired. We had heard a little about this secret warfare, but had never dreamed that we ourselves would play a part in it. It had been a tremendous experience to have been trained by the British Secret Service and to have worked with it.

We had been given first-class training by efficient officers who had completely convinced us of the outstanding qualities of this Service and its leaders, both through their conduct and their complete mastery of the subject. We felt that we had been selected to carry out a task of unparalleled importance!

When faced by incomprehensible contradictions, we laymen did not dare to look upon them as mistakes. And when these mistakes developed, after our arrest by the Germans, into continuous negligence of the grossest kind, we would always rather believe in some mirage born of our despair than give up our blind trust in our Service.

How were we to know that the men who controlled our activities and our fate after we had gone into action would not be of the same efficient quality as those officers who had given us our training?

AUTHOR'S EPILOGUE

WAR is as old as humanity, and as old as either of them is the chequered and adventurous history of Secret Service. Throughout all ages its activities have sometimes prevented, sometimes started and sometimes decided wars, without the history books having been able to chronicle very much about them. A few classical instances have fired contemporary imagination, and have been handed on to succeeding generations, but the great importance and widespread activities of Secret Service in the conduct of war have not, except for occasional accounts of "great spy cases", received any wide publicity.

At the end of every war the secret records of the High Commands of all the nations engaged tend to disappear into the archives of the special departments concerned. These events have no place in the victory announcements; they are suppressed in the official despatches, and the war histories written by General Staffs only make mention of them in so far as the small amount of information and documentary evidence available to the historians makes this possible. This is done for "reasons of State", because media and methods have been in use since time immemorial which are frequently not to be reconciled with the generally accepted code of morals, and not even with the "customs of war". This secret organisation has always and among all nations had its special rules of conduct on account of the particular methods employed—the spy, saboteur or traitor will suffer death, whatever the reason for which he has, in his particular case, taken up this work.

In the German General Staff histories of the First World War, also, only very little is related about secret warfare. The head of the German Secret Service at that time, Oberst i G.

Nicolai, did indeed publish a book about "the secret war". But he has carried his silence beyond the grave concerning his mysterious assistants and their fateful activities.

The records have kept silent, too, since the end of the Second World War. The result is that the public of many countries has been flooded with a mass of baseless legends or deliberately fabricated descriptions of the operations of the Secret Services. The authors have found a grateful and attentive reading public, since these events have lacked a serious presentation, and because silence is still imposed on many individuals who dealt with these affairs at the time.

This account of a portion of my activity in the service of an Abwehr section of the former Oberkommando der Wehrmacht is an attempt to record faithfully from memory some personal experiences from among the events of an apocalyptic period. I have been brought to do this through the idea of making a documentary contribution to the story of the secret warfare waged by both adversaries of the Second World War in Western Europe—a chapter of history about which ill will, ignorance and misunderstanding have started many legends.

I have set aside my original misgivings about the publication of my experiences on account of my desire to reveal the truth about many enigmatical events and occurrences which have closely affected the fate of many Dutchmen.

It is certain that expert historians will some day investigate the origins of this fateful experience of our generation, well clear of the fog caused by the hatreds and prejudice which rack the present. Their best help in this will come from the documentary evidence of eyewitnesses who worked right in the middle of the stream of affairs at this time.

The success of the German military Abwehr in the radio operation *Nordpol* which I started in the spring of 1942 against the English SOE (Special Operations Executive) and which I was responsible for conducting for two years has led to bitter conflicts of opinion in Holland and in England since the end of the war. The investigations of the Dutch Parliamentary Commission (Parlamentare-Enquete-Commissie) have not been able to clear up the question of responsibility for this, the most severe defeat sustained by the Allies in the clandestine war. Ignorance of the true preliminary moves has developed into open imputations of hidden and criminal intentions by the English Secret Service towards the Dutch Underground movement which was carrying on the struggle against the German Occupation.

The tragic consequences of the parallel operation carried out by the German Sicherheitspolizei in Holland and known as the *Englandspiel* contributed towards keeping passions high. The Reichssicherheitshauptamt, which was the authority ultimately responsible for the fate of the *Nordpol* agents, guaranteed to me formally and in writing in the summer of 1942 the physical safety of all the agents who fell into German hands as a result of Operation *Nordpol*. In spite of this, out of fifty-four Dutch-English *Nordpol* agents, forty-seven did not survive the end of the war. The investigations of the Dutch Parliamentary Commission have revealed that they were shot without trial in the autumn of 1944 in Mauthausen camp. Their liquidation was one of the many crimes so typical of the system of the RSHA which cannot be justified by any appeal to the necessities of war. The memory of these victims of an infamous breach of confidence, which I can remember only with shame and bitterness, has guided my pen in the writing of this story.

The concept of clandestine warfare includes espionage, sabotage and the undermining of the enemy's will to resist. Against these forms of attack there is the "defensive Abwehr" by means of special military and political detachments, and secondly the "offensive Abwehr" by means of so-called counter-espionage.

The German counter-espionage service was incorporated at the top in department IIIF of Group III of the Abwehr section of the OKW in Berlin. The title IIIF stands for "Abwehr against foreign services" and applies to the entire military counter-espionage organisation.

Nothing can be written about the activities of the German Military Secret Service in the Second World War without an explanation of the responsibilities, the cleavages and finally the mutual hostility of the different Secret Services of the Third Reich. These differences had a more fundamental basis than simple self-assertion within a common sphere of activity, personal ambition, or the search for positions of influence.

AUTHOR'S EPILOGUE

The great opponent of the Military Secret Service was the Reichssicherheitshauptamt. To it were subordinated the vast organisations of the German security authorities, the STAPO and SIPO, as well as the Sicherheitsdienst and all the other non-military Secret Services. Its head—Himmler—combined unlimited power over this state within a state with command of the divisions of the Waffen SS.

In the RSHA, as well as the political and finally also the military pretensions of the SS, was embodied the claim to leadership in all aspects of spiritual life.

The rejection of these Himmlerite claims by the Wehrmacht was naturally not confined to the military Abwehr alone. Such differences, however, made themselves particularly felt in the Abwehr, where they made fresh appearance daily as a consequence of our close contact with the enemy. This antagonism was particularly dangerous for the Abwehr in view of the efforts of Hitler and Goebbels to awaken "revolutionary militarism" with the object of replacing by officials those officers who were not amenable to the party programmes and the party line.

The attempt to undermine the forces of tradition failed completely in the military Abwehr under the leadership of Admiral Canaris. Canaris earned the implacable hostility of the opportunists of all camps, with the Chief of the OKW, Keitel, at their head, through the steadfastness of his resistance. It brought Canaris to the concentration camp at Flossenburg, where he was murdered on 9th April 1945.

Canaris and many of the men in his service refused to sacrifice their human values to the devilish demands of megalomaniac nationalist pretensions and primitive ideological instincts. Their example should help those of us who have survived to bear the burden of our sad inheritance and to believe in the possibility of throwing a bridge across the flood of abuse, ostracism and proscription which surges round us.

A documentary report must bear the correct name of its author. My work as head of the German Military Counter-espionage in Holland from 1941 to 1944 has become so well-known through pamphlets and novels, press and film, that it would be pointless for me to use a pseudonym. On the other hand, I have made partial use of cover-names or nicknames in the circle of my former assistants.

AUTHOR'S EPILOGUE

By the publication of this book in Dutch I told my story in an "enemy" country whose government was still *de jure* in a state of war with Germany. This circumstance obliged me to observe the greatest veracity and complete objectivity. And this includes my unshakable and never misplaced respect for those Dutchmen who, in a hopeless situation, helped to prepare for their country's liberation by going personally into action in the Secret War.

GLOSSARY OF TERMS

Abwehr	The Secret Military Intelligence Department of the Oberkommando der Wehrmacht (the German High Command).
Afu	Clandestine radio set operated by an agent.
ANR maps	A set of maps issued by the Dutch Cycling Association.
Ast	Abwehrstelle. Local headquarters of the German Military Secret Service.
BBO	1944–5. The Dutch headquarters in London for sabotage and underground armed resistance in Holland.
BI	The headquarters of the Dutch Secret Intelligence Service in London.
Binnenhof	The Dutch Parliament Buildings at The Hague. Occupied during the war by the Chief of the SIPO and the SD (*q.v.*).
Crystal	Means of regulating the wave-length of radio transmissions.
CSVI	Dutch clandestine sabotage and intelligence organisation. Headquarters in Amsterdam. Directed from London.
D/F	Direction finding.
Frontaufklärungs Kommando	Mobile Reconnaissance Units.
FuB station	Radio interception station of the ORPO (*q.v.*).
GFP	Geheime-Feld-Polizei. Secret field security police. Task of GFP was protection of troops and military installations in frontline and occupied territories.
IC	Officer in charge of observation and judgment of enemy force and intentions, working in all military staffs from divisional staff upwards.
IIIC	Military Secret Intelligence Service operating among the civil population.
IIIF	The Military Counter-espionage Department of the Abwehr.

GLOSSARY OF TERMS

Identity check   A plain or concealed indication in the text of radio messages which identifies the sender.

Key-clicks   Radio waves caused by the making and breaking of the keying circuit in a radio transmitter, which can be detected by receivers in the vicinity.

Lauwerszee   The inland sea near the north coast of Holland.

Maréchaussée   The Dutch Gendarmerie.

MID   The Dutch Military Intelligence Section of SOE (*q.v.*) in London.

NSB   Nationaal-Sozialistische-Beweging. The organisation of the Dutch national-socialists. The founder and leader Mussert was executed in the Netherlands after the war.

OKW   The German High Command.

ORDE DIENST (OD)   A widespread underground organisation which operated in Holland during the war. It was composed principally of officials and former officers.

ORPO   Ordnungspolizei, or Regulating Police. The German force responsible in Holland between 1940 and 1945 for the suppression of clandestine radio-links.

OT   Organisation Todt. Nazi organisation for construction and development of big building projects; e.g. autobahns, West Wall, Atlantic Wall, fortifications, airfields, etc.

Radio Orange   The radio transmitter of the Dutch Government in exile in London between 1940 and 1945.

RSHA   Reichssicherheitshauptamt. The headquarters of the Himmlerite Intelligence Services, the STAPO, SIPO and SD.

Security check   Secret indicators in the text of agents' radio messages which confirm to the receiving station that the agent is working in safety.

SD   Sicherheitsdienst. Security Service. An organisation of the Himmler police force for the suppression of "internal resistance".

SIPO   Sicherheitspolizei. Himmler's Security Police. Section IVE of the SIPO was responsible for the liquidation of hostile agents. As services, SIPO and SD worked independently—but both of them were directed by "Befehlshaber der Sicherheitspolizei und des SD" (Commander of SIPO and SD).

SOE   Special Operations Executive. A section of the English Secret Intelligence Service responsible during the war for sabotage and underground armed resistance in the enemy's rear. SOE had special sections corresponding to countries where they operated, for example a Dutch Section for Holland, French Section for France, etc.

STAPO   Staatspolizei or Secret State Police—sometimes known as Gestapo. A Himmlerite organisation.

V-PERSONEN   Vertrauens-Personen. Persons working occasionally or permanently, paid or unpaid, for a secret service.

WA   Afdeeling (Dutch designation). Territorial semi-military formation of Dutch volunteers. Most of them were members of the NSB.

WNVFu   Wehrmacht-Nachrichten-Verkehr-Funk. A department of the OKW responsible for radio intelligence and interception of enemy radio transmissions.

E P
B M
We hope you enjoyed this title
from Echo Point Books & Media

Before Closing this Book, Two Good Things to Know

Buy Direct & Save

Go to www.echopointbooks.com (click "Our Titles" at top or click "For Echo Point Publishing" in the middle) to see our complete list of titles. We publish books on a wide variety of topics—from spirituality to auto repair.

Buy direct and save 10% at www.echopointbooks.com

DISCOUNT CODE: EPBUYER

Make Literary History and Earn $100 Plus Other Goodies Simply for Your Book Recommendation!

At Echo Point Books & Media we specialize in republishing out-of-print books that are united by one essential ingredient: high quality. Do you know of any great books that are no longer actively published? If so, please let us know. If we end up publishing your recommendation, you'll be adding a wee bit to literary culture and a bunch to our publishing efforts.

Here is how we will thank you:

A free copy of the new version of your beloved book that includes acknowledgement of your skill as a sharp book scout.

A free copy of another Echo Point title you like from echopointbooks.com.

And, oh yes, we'll also send you a check for $100.

Since we publish an eclectic list of titles, we're interested in a wide range of books. So please don't be shy if you have obscure tastes or like books with a practical focus. To get a sense of what kind of books we publish, visit us at www.echopointbooks.com.

If you have a book that you think will work for us, send us an email at editorial@echopointbooks.com